TECHNIQUES AND CHALLENGES FOR 300 mm SILICON:
PROCESSING, CHARACTERIZATION, MODELLING AND EQUIPMENT

EUROPEAN MATERIALS RESEARCH SOCIETY SYMPOSIA PROCEEDINGS

TECHNIQUES AND CHALLENGES FOR 300 mm SILICON: PROCESSING, CHARACTERIZATION, MODELLING AND EQUIPMENT

PROCEEDINGS OF SYMPOSIUM F ON
TECHNIQUES AND CHALLENGES FOR 300 mm SILICON
OF THE E-MRS 1998 SPRING CONFERENCE

STRASBOURG, FRANCE, 16-19 JUNE 1998

Edited by

H. RICHTER
Institute for Semiconductor Physics, Franfurt (Oder), Germany

P. WAGNER
Wacker Siltronic AG, Burghausen, Germany

G. RITTER
SEMITOOL Inc, Kalispell, USA

1999

ELSEVIER
AMSTERDAM - LAUSANNE - NEW YORK - OXFORD - SHANNON - SINGAPORE - TOKYO

ELSEVIER SCIENCE Ltd
The Boulevard, Langford Lane
Kidlington, Oxford OX5 1GB, UK

ISBN 0-08-043609-9
Reprinted from: Microelectronic Engineering, vol. 45/2-3

Library of Congress Cataloging in Publication Data
A catalog record from the Library of Congress has been applied for.

British Library Cataloguing in Publication Data
A catalogue record from the British Library has been applied for.

⊗ ,The paper used in this publication meets the requirements of ANSI/NISO Z39.48-1992 (Permanence of Paper).
Printed in The Netherlands.

ELSEVIER

Microelectronic Engineering 45 (1999) v–vi

MICROELECTRONIC ENGINEERING

Contents

Elsevier Science B.V.

ELSEVIER
Microelectronic Engineering 45 (1999) 85–86
MICROELECTRONIC ENGINEERING

Preface

The activities of the semiconductor industry to introduce a new, large wafer diameter were triggered by expected potential overall savings – cost and resource – and an anticipated increasing demand for Silicon wafers. In the beginning, around 1994, agreement on the diameter of the next wafer generation had to be achieved and finally 300 mm was globally accepted to be the next wafer diameter, a decision finally obtained at international summits in 1994/1995, based on the work of a SEMI task force chaired by H. Huff, SEMATECH, and H. Fußstetter, Wacker Siltronic AG and a team of Motorola headed by B. Johnson. The organisations and the companies mentioned were also essential driving forces. Meanwhile all major device manufacturers, respectively, have programs in place to introduce 300 mm manufacturing lines with SC300 being presently in the most progressed position. In addition all silicon suppliers and most equipment manufacturers, in USA, Japan, Europe and Asia, also made good progress with corresponding coordinated activities.

Many obstacles had to be or still are to be surmounted to realize the expected savings. Excellent progress has been achieved in the meantime as is documented by the present conference proceedings volume and most of metrology and process equipment is available for 300 mm wafer processing. However, a lot of work has still to be done and the support of research and development activities by national and international research programs is indispensable and extremely valuable.

Several workshops on 300 mm wafers were held by SEMI, JSNM and other organizations during the past years. However, the present E-MRS conference on "Techniques and Challenges for 300 mm Silicon: Processing, Characterization, Modeling and Equipment" was the first international scientific conference about this subject and we are proud to have had it in Europe. The papers – invited as well as submitted – cover a wide range of subjects, financial issues, fab concepts, crystal growth, wafer process development, material and defect issues, wafer characterization and provide an excellent review of the present status of 300 mm technology.

Organizing an international conference is a tedious and laborious process as everybody knows who undertook such a task. The support by the E-MRS organization and conference team headed by P. Siffert and by the E-MRS conference chairman P. Glasow is therefore highly appreciated as well as the financial support granted by Semitool, SEZ Semiconductor Equipment and Wacker Siltronic AG.

H. Richter,
G. Ritter,
P. Wagner

January 1999

0167-9317/99/$ – see front matter
PII: S0167-9317(99)00139-2

(The conversion process to 300 mm wafers was slowed down by the present business situation in the semiconductor industry. Originally it was thought to happen in parallel with the conversion from 0.25 to 0.18 μm technology generation but most of device manufacturers appear now to aim at a conversion in parallel to the transition to the next, the 0.13 μm technology generation.)

ELSEVIER

Microelectronic Engineering 45 (1999) 87–88

MICROELECTRONIC ENGINEERING

Panel discussion

The 300 mm technology – global opportunity for industry and challenges for research

Chaired by: Dan Tull

Panelists:

Holger Dietrich, Siemens AG, Munich, Germany
Alfred Honold, Jenoptik Infab GmbH, Jena, Germany
Lothar Pfitzner, Fraunhofer Institute for Integrated Circuits, Erlangen, Germany and SEA Dissemination Office, Chilton, UK
Hans Richter, Institute for Semiconductor Physics, Frankfurt (Oder), Germany
Georg Ritter, Semitool Inc., Kalispell, MT, USA
Thomas Theiler, Steag/AST Elektronik, Dornstadt, Germany
Dan Tull, Semiconductor 300 GmbH & Co. KG (Joint Venture Siemens/Motorola), Dresden, Germany
Peter Wagner, Wacker Siltronic AG, Burghausen, Germany

Main topics:

- Conversion of wafer size to 300 mm diameter according to the SIA roadmap is driven by economic factors.
- Equipment manufactures and materials suppliers are preparing the biggest initiative in the history of the semiconductor industry. Extremely high cost and high risk require international alliances.
- Technical challenges concerning material and equipment conversion require cooperation with both the device manufacturers and academic research institutes.
- Among the topics to be studied are metrology problems of large wafers, in situ measurement techniques to reduce the amount of non-productive monitor wafers material problems of the large silicon crystals, and process models for all processes.
- Beside facilities for testing of equipment and material in the USA and Japan, the first pilot line for advanced CMOS devices on 300 mm wafers has now been installed by a Joint Venture of Siemens and Motorola in the new Siemens facility in Dresden, Germany (Semiconductor 300 GmbH & Co. KG).
- 300 mm wafer fabs are characterized by a higher level of process automation, wafer handling, and metrology.

0167-9317/99/$ – see front matter
PII: S0167-9317(99)00138-0

- In addition to the general technology, extremely challenging processes are chemical-mechanical-planarization (CMP), photolithography, dry etching, and furnace processes.
- The major concerns are the economical challenges because at present, the suppliers do not see the possibility of producing new material and equipment for 300 mm wafers meeting the investment cost relation to 200 mm wafers of ≤ 1.3 required by the SEMATECH/I300I model.

ELSEVIER

Microelectronic Engineering 45 (1999) 89

MICROELECTRONIC ENGINEERING

300 mm Conversion challenge and breakthrough for future semiconductor manufacturing*

P. Kuecher, D. Tuft, K. Roithner, M. Hiatt

SEMICONDUCTOR300 GmbH & Co. KG, Germany

Abstract

The status regarding the planned conversion of the semiconductor industry to the larger and more productive 300 mm wafer size is reviewed. Many efforts are already underway to aid the conversion to 300 mm wafers. In addition to the consortiums in both U.S. and Japan, targeted at working with the equipment manufacturers and standard setting, SEMICONDUCTOR300, a joint venture between Motorola and Siemens, will develop a fully integrated process (64M/256M) on a complete toolset. Goal of the SC300 operation is to realize and demonstrate concepts to reduce the cost/die on 300 mm in comparison to 200 mm of approx. 30–40%. A financial model for semiconductor manufacturing using the larger substrates is offered along with the productivity opportunities for each of the equipment types. The authors define equipment manufacturers in terms of performance and financial targets. Achieving these targets is critical to making the conversion happen as well as determining the timing. Finally, the authors offer the breakthrough technologies expected to occur at the 300 mm production level. The noticeable difference between the 300 mm generation of wafer processing in comparison to previous 150 mm and 200 mm conversions is the increase in automation requirements. The 300 mm wafer size will change the way semiconductor factories operate. This presentation offers thoughts of what changes will be required to facilitate 300 mm wafers and beyond. © 1999 Elsevier Science B.V. All rights reserved.

*The main text of this article was not supplied by the authors.

0167-9317/99/$ – see front matter
PII: S0167-9317(99)00252-X

ELSEVIER

Microelectronic Engineering 45 (1999) 91–100

MICROELECTRONIC ENGINEERING

Automation and fab concepts for 300 mm wafer manufacturing

Harald Binder[a,*], Alfred Honold[b]

[a]M + W Zander Facility Engineering GmbH, Lotterbergstrasse 30, D-70499 Stuttgart, Germany
[b]JENOPTIK INFAB GmbH, Goeschwitzer Strasse 25, D-07745 Jena, Germany

Abstract

The introduction of 300 mm wafers into integrated circuit manufacturing will affect the design of the fabs. This paper covers the expectations of IC manufacturers and the progress in technology with respect to the fab itself and automation systems. The fab design is affected by the future use of closed wafer carriers, a consequent automation strategy, new challenges in contamination control, and the increased importance of environmental aspects. It will be described how these factors influence the fab design and the status of the discussed technologies.

Keywords: Si fabrication; 300 mm Si wafers

1. Introduction

The semiconductor industry is well-known for its extremely high innovation rate. The continuous reduction of critical dimensions is combined with increased functionality of integrated circuits (ICs) which results in an increased die size for each new product generation. The number of dies per wafer is considered as one of the critical cost factors. The increased die size forces the IC manufacturers to change the wafer diameter as soon as the number of dies per wafer falls below a critical number. Since 1995, the semiconductor industry has been preparing for the conversion from 200 mm to 300 mm wafers [1]. From the beginning, this conversion was intended to be led not by an individual company as in the past but by consortias, with the consequent promotion of international standards.

The slow down of the semiconductor market led to a delay in investment plans and the shrinking of die sizes within the last few years was more successful than historically experienced. Both effects resulted subsequently in a delay of the 300 mm wafer introduction [2]. 300 mm pilot lines are currently established by some leading IC manufacturers and first 300 mm fabs are expected to be built at the beginning of the next millennium. The conversion to 300 mm wafers creates a new chance to design a new fab generation which places maximum attention to the challenges of contamination control and productivity in IC manufacturing.

*Corresponding author. Tel.: +49-711-8804-1307; fax: +49-711-8804-1950.
E-mail address: harald.binder@mw-zander.com (H. Binder)

0167-9317/99/$ – see front matter
PII: S0167-9317(99)00092-1

2. Expectations of IC manufacturers

The IC manufacturers formed two consortia (one in the U.S. and one in Japan) which tried to direct the supplier industry by establishing general guidelines for this conversion. The guidelines of the International 300 mm Initiative (I300I) gained a lot of attention because they clearly outline the cornerstones of a new manufacturing technology. These guidelines are centered around the goal of an improved productivity and include directions for cleanroom concepts, automation, material consumption, and environmental issues [3].

3. Wafer carriers and equipment load ports

The wafer carrier plays an important role for contamination control as well as automation. I300I decided early to use in 300 mm fabs a closed carrier with an automated door at the front side and a cassette structure which is an integral part of the carrier. This carrier was named Front Opening Unified Pod (FOUP) and forces the process equipment to utilize load ports which accommodate this carrier. It also implies that the process and metrology equipment utilize integrated minienvironments (Figs. 1 and 2).

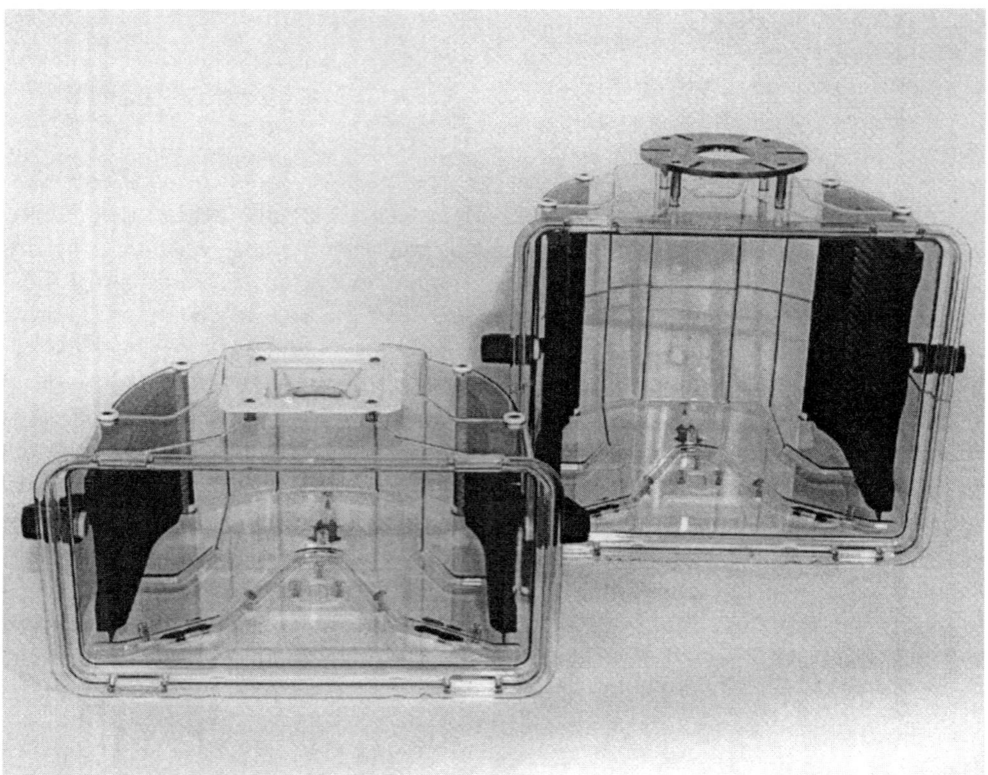

Fig. 1. Front Opening Unified Pod for 300 mm wafers.

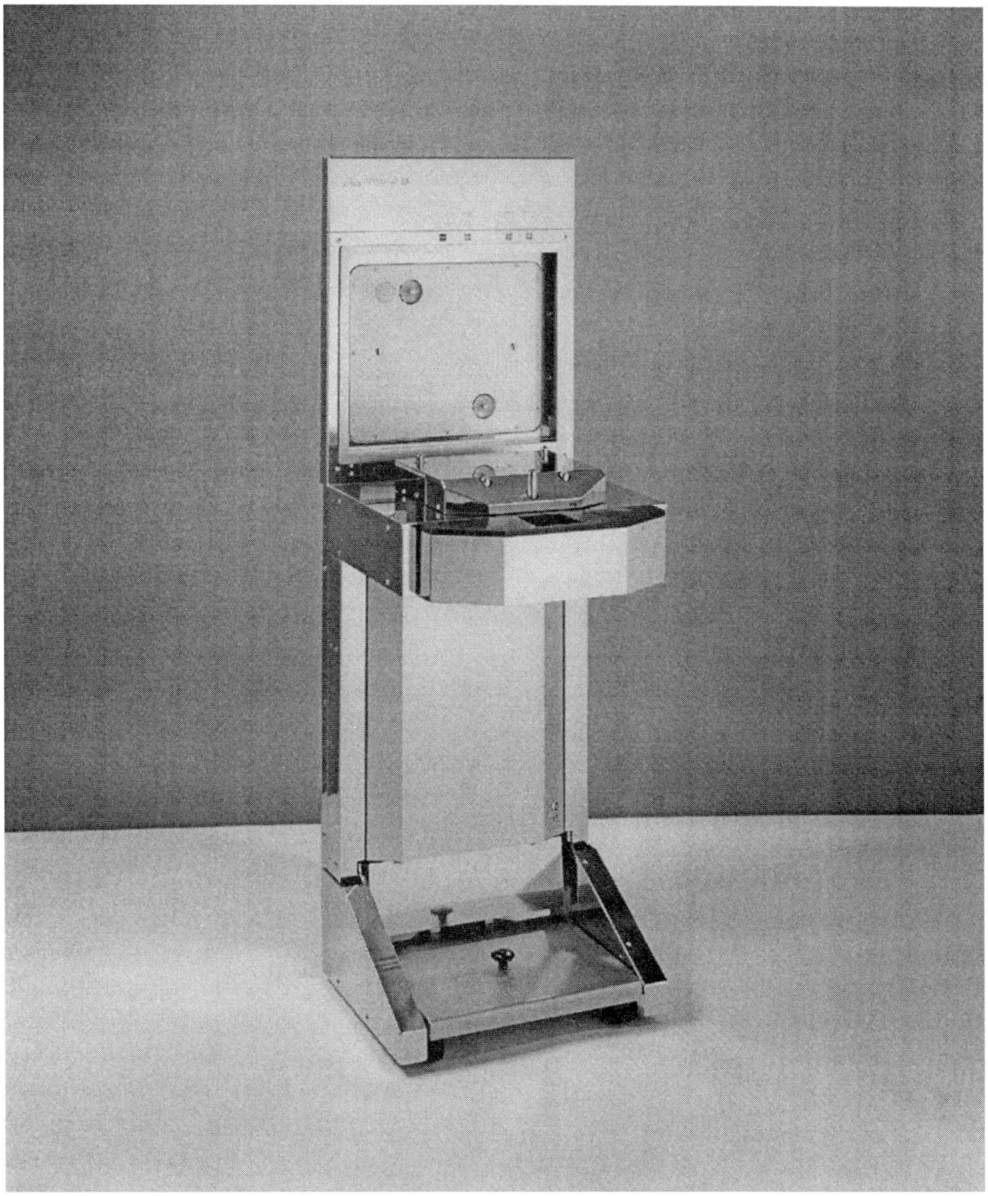

Fig. 2. Load port unit for 300 mm process equipment.

Minienvironments are localized cleanrooms which ensure optimum contamination control in the wafer environment. This technology was introduced ten years ago and demonstrated excellent isolation factors to the ambient which allows to achieve cleanroom classes well below class 1 in an ambient of class 1000 [4]. In addition to particulate contamination, yield loss by molecular contaminants starts to worry IC manufacturers. Minienvironments are a more cost effective way to control the wafer environment (for example via chemical filtering) than traditional cleanrooms.

4. Material transport and storage

The standard configuration for material storage are stockers at the entrance of each process bay. The transport between the stockers is utilized by an automated overhead transport system (OHT). A

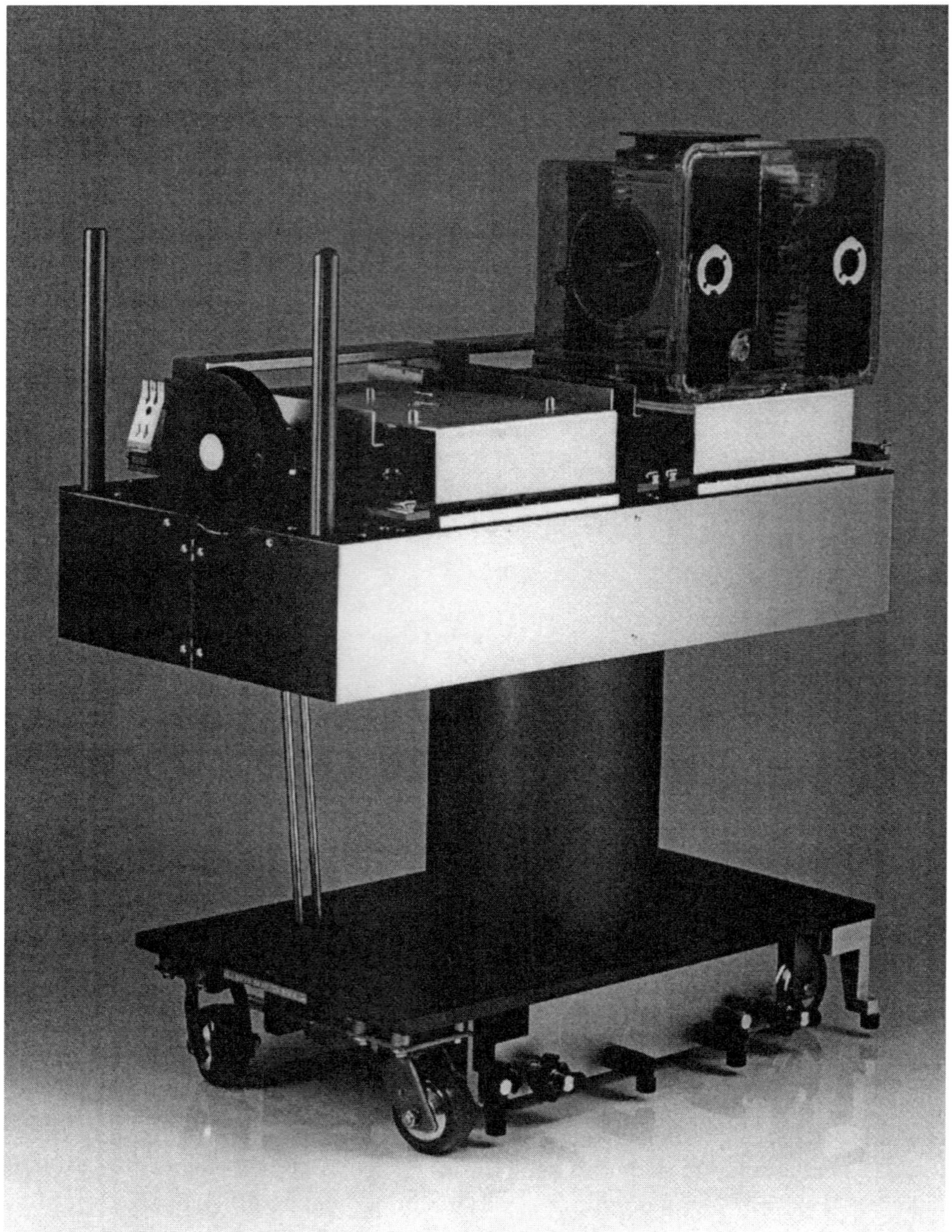

Fig. 3. Personal Guided Vehicle to manually transport and transfer 300 mm FOUPs.

new challenge for 300 mm fabs is the transport within the process bays (intra-bay transport). A FOUP with 25 wafers weighs approximately 8.5 kg which is at the ergonomic limit for operators. In addition, the content of one FOUP represents a very high value and any damage due to handling must be prevented. An automated transport system should not interfere with human operators and must deliver the FOUPs to the equipment load ports which are located at 900 mm height. Automated guided vehicles (AGVs) and rail guided vehicles (RGVs) achieve easily the loading height but operate at the same floor as operators. The OHT does not interfere with the operators but requires sophisticated mechanisms (a hoist system) to transfer the FOUPs vertically between load ports and the transport level just below the cleanroom ceiling. The choice of the IC manufacturer between both alternatives might be driven by criteria such as interference with operators and safety issues. But the biggest challenge behind automated intra-bay transport may be the material control system which has to manage material delivery to particular tools in the right time. The performance of the material control system is expected to affect dramatically the productivity of a fab. The use of local buffers at the tools

Fig. 4. Equipment Front End Module with 4 load ports.

and the application of push instead of pull strategies for material delivery to the process equipment might increase the robustness of the intra-bay material handling system (Fig. 3).

Automated intra-bay transport systems will certainly be the choice for 300 mm fabs in the future but currently they require more development before they can be reliably applied in pilot lines and fabs. An alternative to automated systems are personal guided vehicles (PGVs) which are carts with FOUP transfer mechanisms operated by human beings. It is expected that these PGVs will be used in pilot lines and the first 300 mm fabs and later for the transport of 'hot lots' or as a fall-back solution in case the automated intrabay transport system fails.

One of the I300I requirements is that 300 mm process equipment achieves continuous operation in order to minimize tool idle time. As a consequence, process tools will be equipped with several load ports or even equipment buffers to eliminate any waiting for material to be processed. A good balance between local material storage at the process equipment and central storage in the stockers will be required (Figs. 4 and 5).

Fig. 5. Equipment buffer to store FOUPs at the process tools.

Fig. 6. Minienvironment cleanroom of a 200 mm wafer fab.

5. Cleanroom

The core of a wafer fab is the cleanroom which is required for contamination control. State-of-the-art cleanrooms control particulate contamination at levels of Federal Standard class 1 or below. The focus of contamination control will change from particulate to molecular contamination. Acids, bases, condensables and dopants such as hydrofluoric acid, ammonium, amines, organophosphates or volatile organic contaminants require more attention than in the past due to their interaction with the device itself, the photoresist, stepper optics, process tools, etc. Therefore, control of emission sources and local chemical filtering will be applied in 300 mm fabs.

Due to the use of minienvironment technology, the cleanroom can be reduced to class 100 or 1000. Fan-filter-units (FFU) allow a very cost-effective and flexible cleanroom layout optimized for each process bay. Minienvironments allow even the elimination of cleanroom walls and a move back to ‘open ballroom’ designs. On the other side, the introduction of copper in metallization will require segregation of certain process bays to prevent cross contamination within the cleanroom (Fig. 6).

6. Equipment layout

The typical equipment layout for wafer fabs is the ‘bay-and-chase’ concept. Cleanroom walls separate the process bay from the service chase. Wafers are handled in the process bay which is designed for class 1 or better. The process equipment (except the wafer handling part) is located in the service chase which is operated at class 100 or 1000. Personnel is strictly separated in operators acting in class 1 and service people operating in class 100. The use of minienvironments will make this separation obsolete.

On the other side, the equipment layout is also affected by the wafer transport system, process utilities distribution systems and productivity goals. Basically, the cleanroom footprint should be used as efficiently as possible and transport distances should be minimized. Alternatives to the bay-and-chase layout such as the circular fab, multi-level-fabs or a cellular fab have been discussed within the last few years but they need more time to be elaborated [5].

7. Environmental issues

The I300I Factory Guidelines ask explicitly for serious consideration of environmental issues. Environmental consciousness might still be a 'soft' factor for the success of a company but it has started to become a crucial point to be accepted by the local community in which a fab is to be established. In addition, the cost for the consumption of resources and the waste treatment become important factors in the economic performance of a wafer fab (Figs. 7 and 8).

300 mm process equipment and wafer fabs will be optimized with respect to the consumption of energy, water and chemicals. Methodologies such as mass flow analysis can help to identify the consumption of substances and saving potentials for each tool [6]. In addition, less effluents will be emitted to the environment and water reclaim will become a state-of-the-art technology to minimize the water use. A close cooperation will be required between IC manufacturers, equipment suppliers and fab designers to optimize the environmental performance of a wafer fab. It will also push the innovation in exhaust technology and better separation of chemical waste.

Fig. 7. View of the entire fab including support buildings.

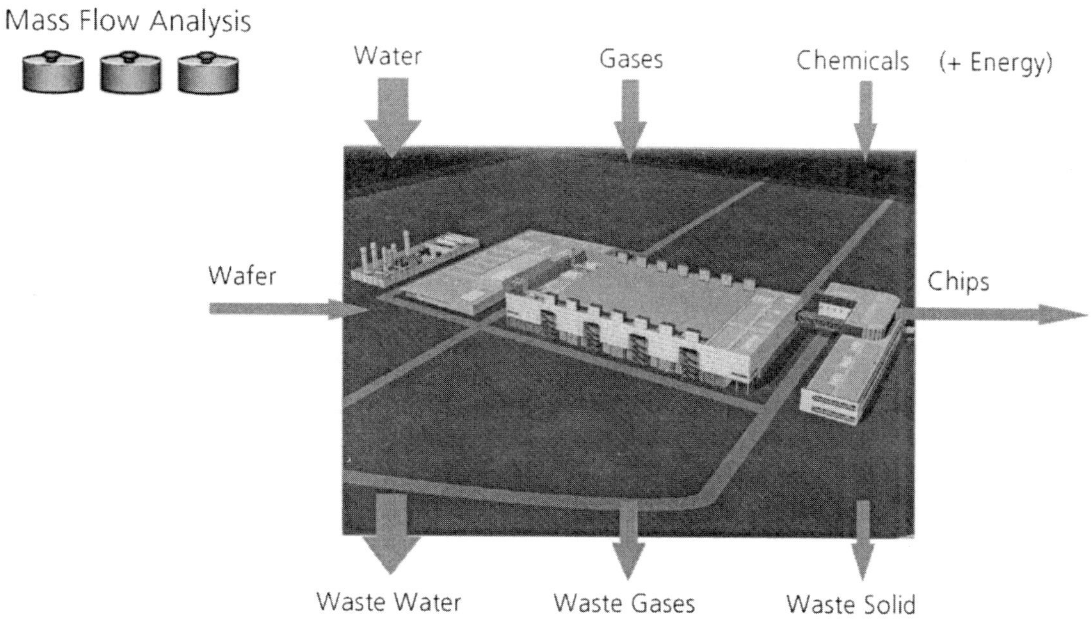

Fig. 8. Methodology of E_4 Efficiency Engineering.

8. Conclusions

300 mm wafer manufacturing will not only push the introduction of new process equipment but requires also a new fab generation. Key parameters in the design of the 300 mm fabs are minienvironment technology, automated wafer transport and storage, control of airborne molecular contamination, equipment layout optimized for productivity and innovation in the control of environmental effects. The expectations of IC manufacturers are high and it will be essential for the success of the entire semiconductor industry to meet these expectations. Even though there are only a few years left to develop the required technologies, it seems that the goals can be met due to efficient cooperation between IC manufacturers, equipment suppliers and fab designers.

Acknowledgements

We would like to acknowledge the input of Peter Csatary (M + W Zander Facility Engineering) and Paul Ballentine (JENOPTIK INFAB) as well as numerous other people in both companies who work on 300 mm issues.

References

[1] F. Robertson, A. Allan, 300 mm Conversion Timing. Future Fab 3, Technology Publishing, London, 1997, p. 27.
[2] F. Robertson, R. Goodall, Semiconductor industry 300 mm conversion business model, in: Proceedings of the Seventh International Symposium on Semiconductor Manufacturing (ISSM), Ultra Clean Society, Tokyo, 1998, p. 53.

[3] I300I Factory Guidelines Version 2.0, Technology Transfer #97063311 B-ENG, I300I, Austin, December 1997.
[4] W. Eissler, H. Schneider, Experimental studies of minienvironment types, in: Proceedings of the 40th IES Annual Technical Meeting, Vol. 1, Institute of Environmental Sciences, Mount Prospect, 1994, p. 396.
[5] K. Potti, Survey of 300 mm wafer fab designs, in: Semiconductor Fabtech, Vol. 6, I.C.G. Publishing, London, 1996, p. 101.
[6] R. Gälli, A. Bachmann, H. Binder, Reduction of production costs through E_4 efficiency engineering, in: Semiconductor Fabtech, Vol. 8, I.C.G. Publishing, London, 1998, p. 85.

ELSEVIER

Microelectronic Engineering 45 (1999) 101–111

MICROELECTRONIC ENGINEERING

Large diameter silicon technology and epitaxy

H. Yamagishi*, M. Kuramoto, Y. Shiraishi, N. Machida, K. Takano, N. Takase, T. Iida, J. Matsubara, H. Minami, M. Imai, K. Takada

Nakanoya 555-1, Annaka, Gunma 555-1, Japan

Abstract

The intermediate results of Super Silicon Crystal project are introduced to serve a production platform of 400 mm silicon. SSi has already installed a 400 mm CZ furnace. Raw materials are sufficiently applicable for 36 inch hot zone. We discuss a contact part between a crystal suspending system and a subsidiary cone. A successful handling system for a fragile quartz crucible and for a heavy silicon crystal is also described. A wire saw for 400 mm wafers has already been installed in SSi and sliced 400 mm wafers are demonstrated. For 400 mm epitaxial wafers, we describe pretreatment for epitaxial growth at low temperature. © 1999 Elsevier Science B.V. All rights reserved.

Keywords: 400 mm silicon crystal; Crucible handling system; Heavy silicon crystals

1. Introduction

DRAM integration density is shifting from 16Mb to 64Mb, however, the predominant diameter of silicon wafers is still 200 mm. However 300 mm silicon wafers are expected to be used around 1999 to manufacture 256Mb DRAM. Samsung announced that 256 Mb DRAM packaging size may be the same as 64Mb DRAM and has proven that the device can be mass-produced from 200 mm wafers rather than having to invest in 300 mm wafer fabrication lines. On the other hand, NEC has announced they will build and equip a 300 mm fab in 2002, expected to produce 256 Mb and 1 Gb DRAM. Therefore the usage of large diameter wafer will be slightly delayed and depends on device maker's shrinkage degree.

The Super Silicon Crystal Research Institute Corporation (SSi) was established in 1996. SSi will develop the key technologies of silicon wafers with a super large-diameter of 400 mm by 2001. It was expected that 400 mm silicon wafers would be in use in 2008. It seems a little early to develop such 400 mm wafers. However we believe that our developing technologies will be useful even for 300 mm wafers fabrication since the platform of 400 mm wafer technology is common to almost every wafer fabrication with a diameter larger than 200 mm.

*Corresponding author. Tel.: +81-27-3840550; fax: +81-27-3840559.
E-mail address: yama@super-si.co.jp (H. Yamagishi)

PII: S0167-9317(99)00106-9

Table 1
Research and Development Targets for SSi's crystal growth and wafer shaping

	Wafer diameter (mm)	Crystal weight (Kg)	Flatness μm)	Particle (μm)	Heavy Metal Impurities (atoms/cm^2)
Targets	400	400	≤0.13	≤0.04	$\leq 10^8$
Present	200	100	≤0.35	≤0.12	$\leq 10^{10}$

In this paper, we mainly focus on SSi's intermediate results of crystal growth and consider a platform to fabricate large diameter wafers.

2. Functions and intermediate roadmap

The targets of our three laboratories, CZ crystal growth, wafer shaping and epitaxial growth are shown in Tables 1 and 2. An intermediate roadmap for research and development of SSi is also shown in Fig. 1.

The first laboratory's targets are to manufacture a CZ crystal growth furnace on an experimental basis and to grow super heavy and dislocation-free silicon single crystal whose diameter and weight are 400 mm and 400 kg, respectively. A first CZ furnace was installed in the SSi laboratory on October in 1997 and a second furnace will be completed in 1999. A 400 mm silicon was supplied to the wafer shaping process at the end of March in 1998.

The second laboratory develops wafer shaping machines on an experimental basis, according to each process such as slicing, lapping, surface grinding, etching, polishing and cleaning. The targets of the wafer quality are shown in Table 1. A wire saw machine was installed in March of 1998 and 400 mm silicon wafers were sliced. Polished silicon wafers with a diameter of 400 mm will be supplied to the epi process at the beginning of the second quarter of 1999.

Table 2
Comparison between 200 mm and 400 mm Epitaxial Wafers

	Diameter 200 mm	Diameter 400 mm
Wafer properties		
Thickness, (micro meter)	725	825
Weight, (g)	52	239
·Warp by self-weight, (micro meter)	23	290
Specification		
Layer thickness (micro meter)	2–5	2–3
Thickness tolerance, (±%)	≤4	≤3
Layer structure defects, (cm^{-2})	≤0.003	≤0.001

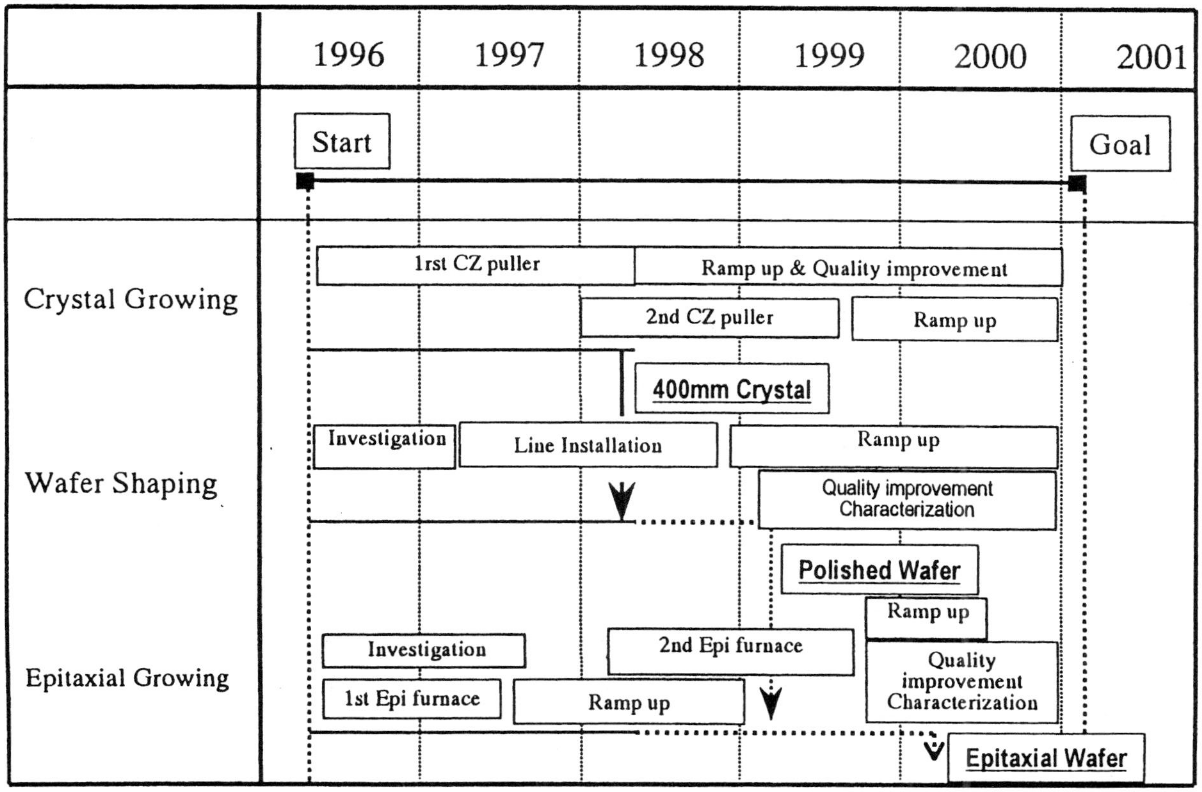

Fig. 1. An intermediate roadmap for Super Silicon project.

The quality requirement both of 200 mm and 400 mm epitaxial wafer is shown in Table 2. Quality requirements, especially impurity level and uniformity, become more severe [1]. SSi installed epi reactor for fundamental experiment such as epitaxial growth at low temperature and pretreatment of wafer surface, in May 1997. Moreover an epi reactor for a single wafer with a diameter of 400 mm was made by MOORE Corp. in the US and installed in August 1997. It is planned that 400 mm epi wafers will be made at the beginning of the second quarter of 2000.

3. Developed key technologies for CZ crystal growth

The following key technologies must be developed to achieve these targets for CZ crystal growth; CZ crystal growing furnaces with a crystal suspending system, large hot zone parts, handling and transportation system for heavy hot zone parts, retrieving system for a grown crystal from the CZ furnace, spherical grinding machine on a grown crystal and ingot cutting band saw. We also consider an anti-steam-explosion safety in CZ furnace and crystalline quality. These targets are described in details in [2–4].

Here we focus on our installed equipment.

3.1. CZ crystal growing furnace

SSi installed the first huge CZ crystal growing furnace, made in collaboration with Leybold Systems Corp. in Germany, in the end of 1997 and started crystal growing in the beginning of March in 1998. Its total height from the ground level and total weight are about 12 m and heavier than 30 metric ton respectively. A super conductive cusp type magnet made by Mitsubishi Electric Co. is provided surrounding the first CZ furnace to suppress silicon melt convection and to control interstitial oxygen concentration. Its outer diameter, inner diameter, height and weight are about 3 mm, 2 m, 3 m and 10 metric ton respectively. The maximum magnetic field at the 36 inch quartz wall is 3000 gauss. Its storage energy is 9 MegaJ, which is very much higher than that ordinary used super conductive magnet of 1 MegaJ. It is controlled from an isolated operating room to avoid steam explosion and straying magnetic field. Its present status is in alternate improvement for initial troubles and crystal growing using 36 inch hot zone.

The first grown silicon crystal whose body length and weight are 400 mm and 180 kg respectively is shown in Photo 1, comparing with a 300 mm silicon single crystal. Our present main target is dislocation free single crystal growth by the end of 1998.

3.2. Hot zone parts and quartz crucible

Isotropic graphite material and carbon composites (CC) are supplied by TOYOTANSO Co. and NIHON CARBON Co., respectively. Their size is available for a hot zone size of 102 cm (40 inch) in diameter. We are now using hot zone size of 91.4 cm (36 inch) in diameter without any problems.

Quartz crucibles with a diameter of 36 inch were supplied by Shin-Etsu Quartz Products Corp. and Mitsubishi Material Corp. Slight desformation of the quartz crucible is observed after CZ crystal growth. It is not obvious whether the corrosion of quartz crucible's surface induces the structure loss during CZ crystal growth. Takano et al., predicted the thermal environment in CZ furnaces for silicon crystals having a diameter of 15, 200, 300 and 400 mm by numerically global heat transfer simulation [5]. When we improve the 36 inch hot zone structure for silicon crystal with a diameter of 400 mm, the heater power and maximum temperature on the quartz crucible are almost the same as those of 24 inch hot zone for 200 mm silicon crystal growth. Moreover dislocation-free crystal growth is observed even after re-melting process due to structure loss phenomena. Therefore we believe it possible to obtain dislocation free 400 mm single crystals by improving SiO particles in the 36 inch hot zone.

3.3. Handling and transportation system for heavy hot zone parts and crystals

In CZ crystal growth process, we have to handle a quartz crucible in the weight region of 70 kg. First, the contamination on its inner surface is removed by cleaning by slashing liquid through a nozzle inserted from the beneath of the crucible. Subsequently, we have to turn over it and to insert into a graphite crucible. The quartz crucible is so fragile that we have to sustain it by stressing appropriate pressure at both sides of the crucible. We have designed several attachments for handling to turn over and insert it into a graphite crucible, automatically.

A grown crystal is retrieved from the CZ furnace. LS's CZ furnace provides a crystal removing system from the growth axis. We developed a crystal retrieving equipment from LS's removing equipment. We utilized a crane that runs along two dimensionally under the ceiling. Usually, a

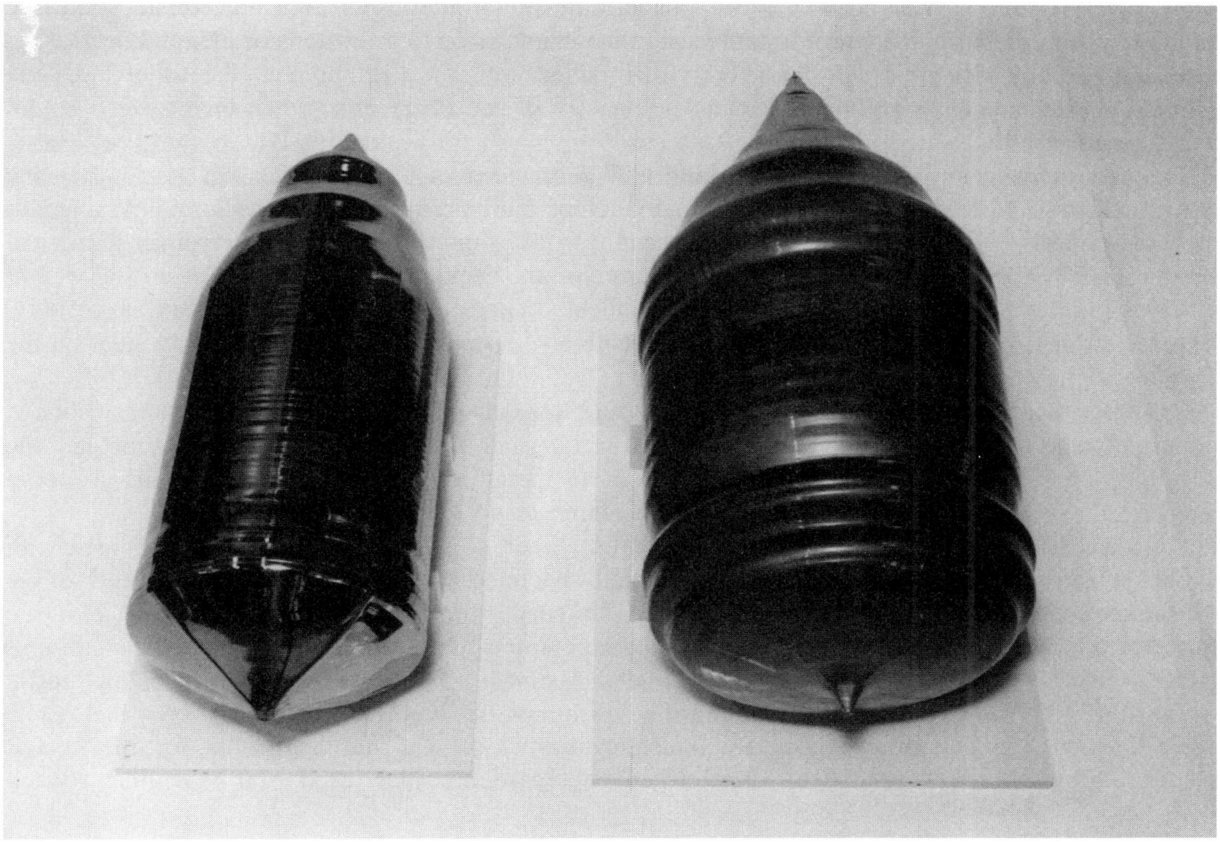

Photo 1. SSi project's first 400 mm silicon crystal. A 300 mm single crystal is shown for reference. Each body length: 400 mm.

retrieving car is used, running on a floor made of stainless steel to avoid magnetic interacting forces between the super conductive magnet and the floor. However if we suppose a crystal heavier than 400 kg, the weight of the car becomes several metric tons, which requires very a expensive thick stainless steel floor. Therefore our equipment is suitable for the MCZ system.

3.4. Crystal suspending system

The weight of a grown 400 mm silicon single crystal becomes heavier than 400 kg. Therefore a minimum necking diameter of 6 mm for the crystal weight of 400 kg is needed. Recently Chandrasekhar reported a a necking success with a diameter larger than 10 mm in high boron doped silicon crystal growth [6]. However we believe that it is not easy to grow such a necking in heavily charged silicon melt since we have to control the melt temperature to maintain the curvature of the growth interface. Moreover we have to consider a safety factor larger than 2 to grow a single crystal safely with creep phenomena at high temperature around 800°C.

Therefore we have to invent a suspending system of growing silicon single crystals instead of necking [2–4]. During necking process, its diameter is kept around 3 mm to achieve dislocation free and then a subsidiary cone is made, by increasing the diameter up to a few tens of mm and decreasing again up to about 10 mm. The diameter is maintained until the beginning of the following cone process. A mechanical cramp is attached to the beneath of subsidiary cone when the growing crystal weight increases to about 150 kg.

The maximum tensile stress in the necking with a diameter of 15 mm is 2.7 kgf/mm^2 when the crystal weight is 400 kg. Takano et al., calculated temperature profile along the necking with a length of 1 m when the body length increases from 40 mm to 1170 mm by a global heat transfer numerical simulation [7]. The calculated maximum temperature on the necking is 600°C when the crystal weight is 400 kg. After Yasutake's result [8], the mechanical strength of silicon between 700°C and 500°C increases from 10 kgf/mm^2 to 64 kgf/mm^2, respectively. Therefore we can theoretically suspend the crystal weight of 400 kg in the safety factor of 4.

For the actual process, the attaching area between the mechanical cramp and the subsidiary cone is important so as not to induce slip dislocation and cracking into the growing necking. We simulated the attaching problem using a tension test at room temperature. We describe the results at room temperature since a large tensile stress testing machine of up to 800°C is unavailable.

We made two types of specimens for the testing, silicon or stainless steel necking provided with double subsidiary cones at both edges. We also made specimen holders attaching to the beneath of the subsidiary cones, by using Mo or stainless steel. The tensile stress curves are shown in Fig. 2. In the case of stainless specimen, every attachment material exhibits non destructive up to a tensile stress of 800 kg. However, in the case of silicon specimens, a silicon specimen was destroyed below tensile stress of 400 kg for a metal attachment and the attachment of CC itself was destroyed at the tensile stress of 600 kg. Therefore non metal attachments such as CC are applicable for the crystal suspending system. However CC attachment generates particles since the crystal habit lines scratch

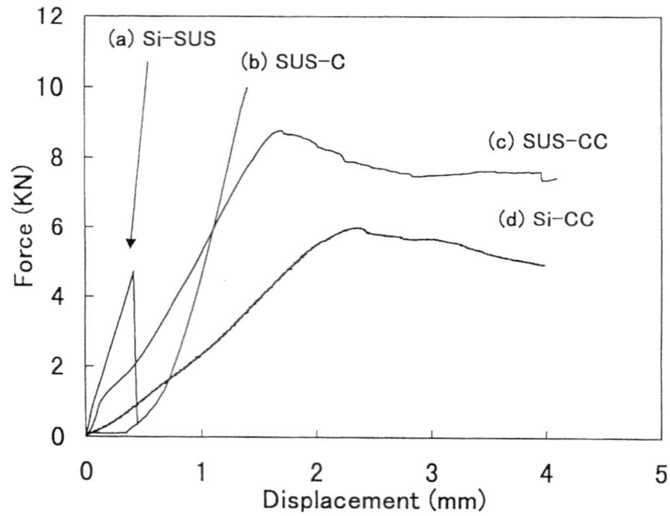

Fig. 2. Result of tension test for a contacting part between the underneath of a subsidiary cone and a crystal suspending attachment. Each curve is corresponding to a combination between a subsidiary cone made of Si or of stainless steel (SUS), and the attachment made of graphite (C), SUS or carbon composites (CC).

the surface of the CC one. Moreover the safety factor is 1.5 for the suspending weight of 400 kg, which is insufficient for our targets. Therefore we need further investigation about the contacting part.

4. Key technologies for wafer shaping

The followings are key technologies to be developed to achieve the targets of the wafer technology laboratory [4]; wafer slicing by multi-wire saw, wafer flattening by surface grinding, double side polishing, double disc, single wafer cleaning and wafer evaluation technologies. Here we describe our recent result of a wire saw [9] and a high rigid surface grinding [10].

4.1. Wafer slicing by multi-wire saw

SSi's wire saw (made by: TOKYO SEIMITSU, Japan), schematically drawn in Fig. 3, was designed so that the various parameters were more widely variable than the standard machine for the purposes of testing the optimum slicing conditions for 400 mm wafers, and of collecting the basic experimental data on wire-saw slicing in general. The SSi's wire saw can slice up to 400 mm in diameter, 400 mm in length, and 120 kg in weight. Its wire tension and wire speed are available up to 7 kgf/wire and up to 1000 m/min, respectively, which are two or three times as large as those of the conventional machine. The machine has adopted the specially designed small sized-high power driving motor and the high capacity main spindles for the groove rollers. The tension separation mechanism between the groove rollers and the wire spools has also been adopted.

Several sliced wafers by using our wire saw are shown in Photo 2. Although we used a dislocated

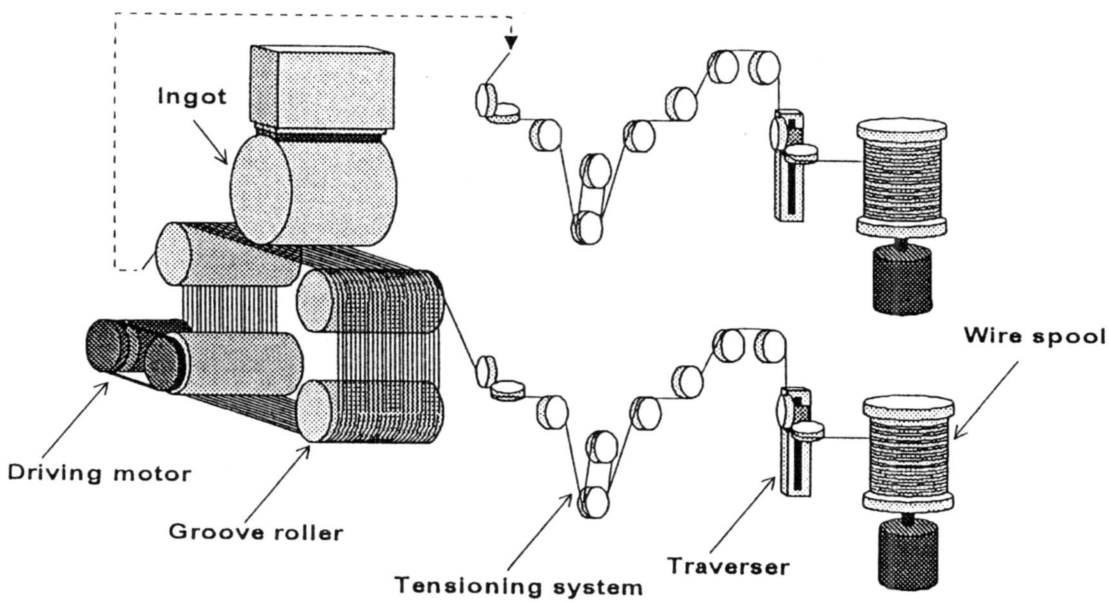

Fig. 3. A schematic drawing of a wire saw.

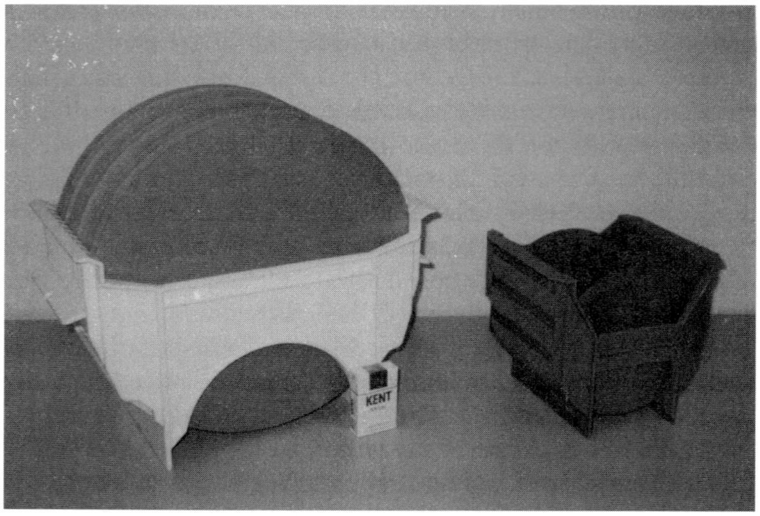

Photo 2. Two 400 mm silicon wafers. 200 mm wafers are also shown for reference.

silicon ingot for the slicing, these wafers are sliced very well without hard warpage and cracks. We will exhibit these wafers in SEMICON WEST that is scheduled for July 1998.

4.2. High rigid surface grinder

Although immersion type etching has been used following the lapping in the standard 200 mm wafer process, we are trying to replace the etching with surface grinding in the SSi's project. The etching process is to remove the damaged layer caused by lapping and to make the satin finishing backside of a wafer. However, as the backside of 400 mm wafers requires mirror finishing, there is no more necessity to do the etching. It is possible to construct an etching-less wafer shaping process by introducing the shallow damaged surface grinding technology.

The motion copying-surface grinding is attractive to realize the high flatness of a wafer and to maintain its precision. In designing the surface grinder (made by: DISCO, Japan), we have adopted an infeed mechanism with nanometer accuracy and strengthened the spindle rigidity by a large margin. By these improvements, we are aiming at a flatness target of 0.13 μm steadily.

However, as for a practical use of the ductile mode grinding in the next generation, there are still several problems left. Namely, in order to adopt the ductile mode grinding in the mass production level, the appropriate tool (grinding wheel) technology matching to the high accuracy of the surface grinder and the proper dressing technology of the tool needs to be established.

In SEMICON-WEST 1998, we will exhibit 400 mm wafers ground by the machine.

5. Key technologies for epitaxial growth

Technical issues for 400 mm epitaxial wafer development are summarized to following categories [4].

1. Decrease of metal contamination on surface and in epitaxial layer.
2. Reduction of production cost.
3. Decrease of defects generated from particles during epitaxial growth.
4. Prevention of slip generation.
5. Wafer support method.
6. Uniform wafer heating.
7. Improvement of Gettering capability.
8. Improvement in thickness deviation over whole wafer.
9. Thinner epitaxial layer and narrower (steeper) transition region.

SSi is conducting the following research for epitaxial growth processes; low temperature process, Cl gas compound free process, analysis of epitaxial process by computer simulation and employment of process tailored substrates.

Here we introduce development of low temperature process and the other issues are written in Ref. [11].

5.1. Development of low temperature process

Lowering of process temperatures is an effective means to decrease contamination, prevent slip generation, and lower cost. We are developing a process below 1000°C. Possibly the largest breakthrough technology on the agenda, such lower process temperature will achieve clean surface before epitaxial growth.

We have developed basic experiment equipment, expressly for the purpose of researching relationship between surface cleanness and epitaxial process conditions. This equipment is a combination of XPS and CVD processes; samples are transported between processes in a vacuum as shown schematically in Fig. 4. XPS was improved to facilitate measurements for total reflection conditions [12,13]. Improvements in the detection limit by two orders of magnitude are expected, allowing oxygen and carbon compound detection to the level of 10^{11} atoms/cm^2.

We are performing to measure impurities on wafer surface after wafer cleaning and heat-treatment. We will decide optimum surface condition for epitaxial substrate and heat-treatment conditions before low temperature epitaxial growth.

6. Conclusion

We have introduced SSi's targets and discuss key technologies. A CZ furnace for 400 mm silicon crystal growth has already installed. A crystal suspending system should be developed since the weight of CZ grown silicon single crystal becomes heavier than 400 kg. We discuss the contacting part between a subsidiary necking and a suspending system attachment. Raw materials such as isotropic graphite are sufficiently available for 400 mm crystal growth. For the wafer shaping process, we have shown our wire saw and demonstrate sliced 400 mm wafers. A 400 mm epitaxial silicon wafer grown at relatively low temperature may be dominant for giga-bit era.

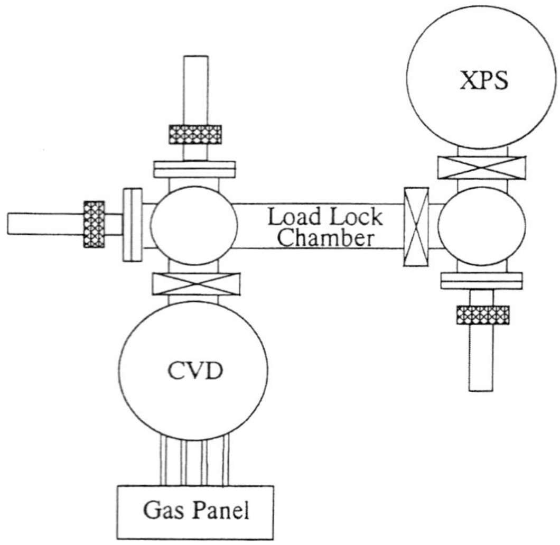

Fig. 4. A schematic drawing of an experimental equipment for epitaxial growth at low temperature.

References

[1] The National Tech. Road Map for Semiconductors, Semiconductor Industry Association, US, 1997.

[2] H. Yamagishi, M. Kuramoto, Y. Shiraishi, N. Machida, K. Takano, N. Takase, T. Iida, J. Matsubara, K. Takada, CZ crystal growth development in super silicon project, in: M. Umeno (Ed.), The 145th Committee of The Japan Society for The Promotion of Science, Tokyo, Proceedings of The 2nd International Symposium on Advanced Science and Technology of Silicon Materials, 1996, p. 59.

[3] H. Yamagishi, M. Kuramoto, Y. Shiraishi, N. Machida, K. Takano, N. Takase, T. Iida, J. Matsubara, K. Takada, CZ crystal growth development in super silicon project, Gettering and Defect Engineering in Semiconductor Technology 1997, in: C. Claeys, J. Vanhellemont, H. Richter, M. Kittler (Eds.), V57-58 of Solid State Phenomena, Scitec Publications, Switzerland, 1997, p. 37.

[4] K. Takada, H. Yamagishi, H. Minami, M. Imai, Research and development of super silicon wafers, Semiconductor Silicon 1988, in: H.R. Huff, H. Tsuya, U. Gosele (Eds.), The Electrochemical Proceeding Series, 1998, p. 376, PV-98-1.

[5] K. Takano, T. Iida, M. Kramoto, N. Machida, J. Matsubara, Y. Shiraishi, N. Takase, H. Yamagishi, 'Thermal environment prediction for silicon crystal growth of up to 16-inch diameters in Czochralski furnaces', Jpn. J. Appl. Phys. 36 (1997) 1563.

[6] S. Chandrasekhar, K.M. Kim, 'Growth of large diameter (> 10 mm) necks for large size CZ silicon', Semiconductor Silicon 1988, in: H.R. Huff, H. Tsuya, U. Gosele (Eds.), The Electrochemical Proceeding Series, 1998, p. 411.

[7] K. Takano, Y. Shiraishi, T. Iida, N. Takase, J. Matsubara, N. Machida, M. Kuramoto, H. Yamagishi, in: Ext. Abstr. of 58th Autumn meet./1997, Jpn. Society Appl. Phys. and Related Societies, 4a-T-4.

[8] K. Yasutake, J. Murakami, Mechanical properties of heat-treated CZ-Si wafers from brittle to ductile temperature range, Jpn. J. Appl. Phys. 21 (1982) L288.

[9] H. Ohishi, K. Asakawa, J. Matsuzaki, Design concept of multi-wire saw for 400 mm diameter silicon ingot slicing, in: Proceedings of the Silicon Machining 1998 Spring Topical Meeting, The American Soc. for Precision Engineering, 1998, p. 109.

[10] K. Abe, S. Ohkawa, Y. Koma, K. Arai, Development of an ultraprecision grinding machine for super-large and super-flat silicon wafer, in: Proc. of the Silicon Machining 1998 Spring Topical Meeting, The American Soc. for Precision Engineering, 1998, p. 113.

[11] M. Imai, M. Mayusumi, K. Inoue, S. Nakahara, The change of surface micro-roughness on epitaxial growth process, in: Meeting Abstracts of The 193rd Meeting of The Electrochemical Soc., 1998, Abstract No. 508.
[12] M.J. Chester, T. Jach, S. Thurgate, Grazing-incidence X-ray photoemission spectroscopy investigation of ixidized GaAs(100): A novel approach to nondestructive depth profiling, J. Vac. Sci. Technol. B11 (4) (1993) 1609.
[13] M.J. Chester, Grazing incidence X-ray photoelectron spectroscopy from multilayer media: Oxidized GaAs(100) as a case study, Physical Review B 48 (23) (1993) 17262.

ELSEVIER

Microelectronic Engineering 45 (1999) 113–125

MICROELECTRONIC ENGINEERING

Challenges for economical growth of high quality 300 mm CZ Si crystals

Erich Tomzig*, Wilfried von Ammon, Erich Dornberger, Ulrich Lambert, Werner Zulehner

Wacker Siltronic AG, D-84479 Burghausen, P.O. Box 1140, Germany

Abstract

The changeover from 200 mm to 300 mm is required by the semiconductor industry due to the necessity for larger chip sizes and demand for decreasing cost. However, the cost for 300 mm crystal growth is likely to rise owing to larger puller, enlargement of hot zone, expensive silica crucibles and longer growth process times caused by lower growth rates and longer cooling rates. Simultaneously, the conditions are more complex and disadvantageous to the required higher qualities in comparison to smaller wafer diameters (e. g. position of OSF ring). This paper gives an overview about the challenges for 300 mm growth and approaches to provide appropriate solutions (e.g. application of magnetic systems, optimization of growth parameters by integration of numerical simulation).

Keywords: Si crystal growth; 300 mm wafer; Magnetic systems

1. Introduction

Since the beginning of IC-technology the chip size has been growing caused by the higher complexity of every new IC generation. Shrinking design rules are not sufficient to compensate higher integration of functions and a larger chip size reduces the number of chips per wafer resulting in higher costs. Larger wafer diameters offer the only alternative to increase chip yield. As a consequence, the wafer diameter has been increasing along with the development of chip size. Meanwhile, the changeover from 200 to 300 mm has been decided for the next generation by the semiconductor industry. It increases the wafer area by a factor of 2.25 and the utilizable area by about 20%. The expected cost reduction is between 20 and 30% [1].

The transition from 150 to 200 mm wafer promised a reduction of more than 20% in the costs per die, but resulted in less than 15% [2]. Regarding the further increase of wafer size, however, it is not clear whether or not the wafer manufacturer will be able to reduce the costs per sq. cm polished wafer. After first estimations the cost of silicon will become an increasing percentage of the fabrication process cost per sq. cm by the changeover to 300 mm [1]. The major reason for this situation is the

*Corresponding author. Tel.: +49-8677-83-2084; fax: +49-8677-83-7303.
E-mail address: erich.tomzig@wacker.de (E. Tomzig)

0167-9317/99/$ – see front matter
PII: S0167-9317(99)00109-4

fact that the upscaling of crystal growth processes for large diameters causes severe problems in keeping an acceptable economy and, at the same time, providing the required bulk quality.

In this paper the essential problems are presented which lead to this situation. Concepts are discussed to solve these. It has to be noted that the next transition to 400 or, as discussed, to 450 mm dramatically intensifies the challenges and, as a consequence, the cost situation.

2. Challenges for 300 mm crystal growth

2.1. Required charge weights

With increasing crystal diameter, the economical conditions become worse by

- lower pull rates,
- longer melting and cool down phases,
- longer growing times for seed and tail end cones,
- lower ratio between the usable cylindrical length of crystal and unusable material (cones, residual melt),
- higher material losses in case of dislocations.

The last point demands an absolutely perfect growth process to avoid dislocations. Due to the corrosion problems and, as a consequence, the limited lifetime of silica crucibles the remelting of a crystal after structure loss is rather unfavorable, because it usually leads to low or zero yield.

In order to compensate the points listed above, it is necessary to increase the charge weight from presently 100 kg in the 200 mm production up to 300 kg or more to obtain economical length and throughput for 300 mm crystal growth. Accordingly, the size of the hot zone which is characterized by the silica crucible has to be enlarged from 22″–24″ to 32″–36″.

2.2. 300 mm puller

For shortening the 300 mm growth development and collecting experience we have started our 300 mm activities in existing pullers, originally designed for 200 mm production, as done by other competitors, too. After a phase of growing 300 mm crystals with 24″ hot zones we have increased the diameter to 28″ by using carbon composite (CC) parts. These allow to reduce the wall thickness of hot zone parts in comparison to the conventional graphite parts. The 28″ hot zone allows to use charge weights of 180 kg, yielding a usable crystal length of about 75 cm.

For further enlargement of the hot zone it was necessary to design a new puller generation. First studies have shown that an upscaling of the traditional layout for the shaft puller does not fulfill the requirements, e.g. handling, stiffness, axiality and absence of vibration, which are necessary for large crystal weights. Therefore, the conventional concept with one hydraulic column was dropped and a 3 column layout was developed by Leybold AG, our supplier of the first 300 mm puller. Because of the large diameters and high weights of hot zone parts, charged crucibles, and grown crystals, the handling was fully integrated into the equipment design. Multiple safety systems were considered for the case of a steam explosion. The hot zone was especially designed, so that in case of a broken silica crucible no liquid silicon can get in contact and react with the chamber walls. In view of mass production, the puller is equipped in such a way that the operators can stay in the hall for controlling and recharging the pullers every time in contrast to the reported research puller.

In order to reduce cost for designing and manufacturing the puller it was intended to develop a standard 300 mm puller for the market. But the individual requirements of each silicon producer caused by e.g. hot zones, crucible diameters (32″, 36″ and 40″), magnetic systems and optional components differ too much to realize a standard design. So, all 300 mm pullers, manufactured by Leybold for different customers, show large differences and only have in common the developed basic 3 column design [4] (see also [5]).

It must be mentioned that the cost and complexity of the puller depends also on additionally required components and options, e.g. additional or movable heaters, movable hot zone parts, magnetic and suspension systems.

For the 300 mm growth, other equipment suppliers have designed cable pullers, e.g. Kayex [6]. The advantage of this puller type is that the cable provides a self-adjusting system, which always gives a right angle. Therefore, only one column is needed for movement and handling of the different chamber parts. But the question is open, which type of puller — shaft or cable type — will be more suitable for growing large diameter crystals. Because of the risk of orbiting of seed and crystal by using cable type the success may be strongly dependent on hot zone design and used growth parameters.

2.3. Magnetic systems

The application of magnetic systems seems to be favorable or even necessary for large melt volumes to damp and control melt convection. The presently used types of magnets, consisting of vertical, cusped or horizontal coils lead to

- reduction in oxygen content in the grown crystal,
- reduction of the temperature fluctuations (e.g. [7–9]) caused by the buoyancy driven turbulent flow, which are typically higher in large melt volumes,
- reduced corrosion of silica crucibles due to slowing down of melt convection and, as a consequence, to lowering the dissolution rate or
- reduced influence of corrosion, because the SiO_2 particles which are emitted into the melt due to corrosion are dissolved before they arrive at the growth interface.

The last listed advantages and the reduction of temperature fluctuations may be very helpful for a robust 300 mm growth. But the fact of oxygen reduction by magnetic fields and the large crucible diameter may lead to problems if a higher oxygen content in the crystals should be specified in the future.

However, the purchase and the use of magnetic fields are a considerable cost factor, e.g. cusp magnets for the needed equipment diameter and field strength of about 0.15 T have a weight of more than 10 tons and a power consumption of more than 400 kW. Even magnetic fields with field strengths up to 20 000 Gauss will be tested for large diameter growth [10]. Additionally, sufficient and expensive screening of the radiation is necessary for application in mass production. The operating cost can be decreased by using superconducting magnets.

2.4. Necking

The conventionally used Dash technique for necking is no longer sufficient for growing large ingot weights. Besides the problems of creep phenomena [3,11], the diameter of a 5 mm grown neck is not able to support crystals of more than 250 kg. Our tests have confirmed the published data [12,13]. The necks with diameter between 3.8 and 4.8 mm broke at a load in the range of 260 to 380 Nm at room temperature.

Suspension systems, which hold the growing crystal for instance at a subsidiary cone [14] might be an alternative, but lead to more mechanical and control complexity.

An alternative necking method, where a reduced pull rate was applied during necking, promises dislocation-free necks of diameters up to 12 mm, in the case of heavy boron-doping [15]. Another publication describes successful necking with a seed having a pencil-like shape of tip, which reduces the thermal shock during dip in [16]. The latter two methods are less complex than a suspension system. It remains to be verified which method will have the best capability in a production line or whether alternative methods are required.

2.5. Consumables

2.5.1. Silica crucibles

Silica crucibles play a key role for the cost of the crystal growth. They are the second highest cost factor next to silicon poly material and their quality is a dominant factor for crystal yield. Even in 200 mm growth, yield is strongly dependent on the 'small' diameter (22″ or 24″) silica crucible qualities. Detailed investigations of used silica crucibles have revealed the following impacts (Fig. 1):

- melt down and growth times, which are longer for large charge weights,
- crucible wall temperatures,
- nuclei for cristobalite growth, e.g. due to surface impurities, which cause pitting corrosion and, as a result, the generation of small particles, which can cause structure loss at the growing interface,
- large gas bubbles or clusters of bubbles near the inner crucible surface. If the silica layer above these is dissolved, they burst open and emit small particles, which can generate dislocations.

Further improvements of crucible quality by reduction of surface nuclei and bubble size, density and growth are necessary and, as shown in the past, possible.

One alternative may be the passivation of the inner crucible surface [17]. However such crucibles are not commercially available for all silicon producers. Moreover, as an alternative, applying magnetic fields reduces the corrosion of crucibles as mentioned above.

Today the first silica crucibles with an inner layer of synthetic silica instead of natural silica glass are commercially available. Our tests have shown that the delivered qualities are strongly varying. But it is reported [18] that longer lifetime was obtained than with silica crucibles consisting of natural silica.

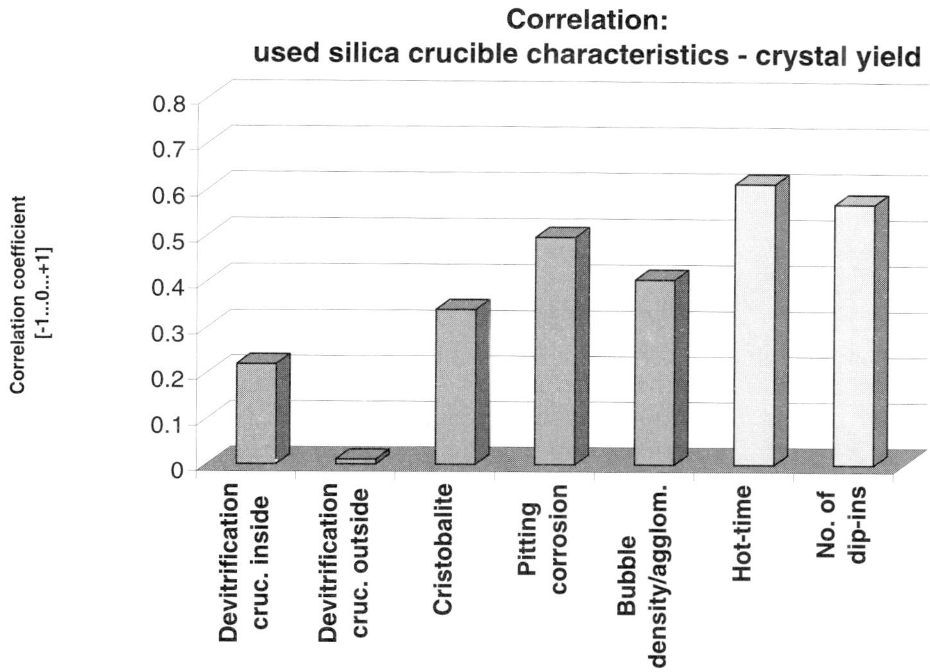

Fig. 1. Correlations between different characteristics of used silica crucibles and the achieved yield.

It remains to be seen whether or not the quality and at the same time the cost of silica crucibles can be improved according to the requirements.

2.6. Graphite consumables

For equipping large 300 mm pullers with hot zone parts, the development of large diameter graphite blocks was necessary. Now the technique is available for isostatic pressing of blocks with sufficiently large diameters. At the same time, the graphite quality was improved. Material with higher density and smaller porosity, which have a positive impact on the lifetime, is available. Further improvement is ongoing.

Additionally, carbon composite (CC) material has been tested for hot zone parts. This material offers the advantage of high strength and low porosity. Thin wall thicknesses are possible, which have a positive impact on the weight of the large hot zone parts and reduce the time constants for process control due to the reduced mass. One-piece susceptors prevent liquid silicon from leaking out into the growth chamber if the silica crucible breaks. The high cost for such material might be compensated by longer lifetimes compared to conventional graphite. The CC development is in progress especially for suitable designing and manufacturing different hot zone parts.

2.7. Process development

In the past, the development of new growth processes was very time consuming. For each specification, an improved parameter set must be found by performing series of growth experiments.

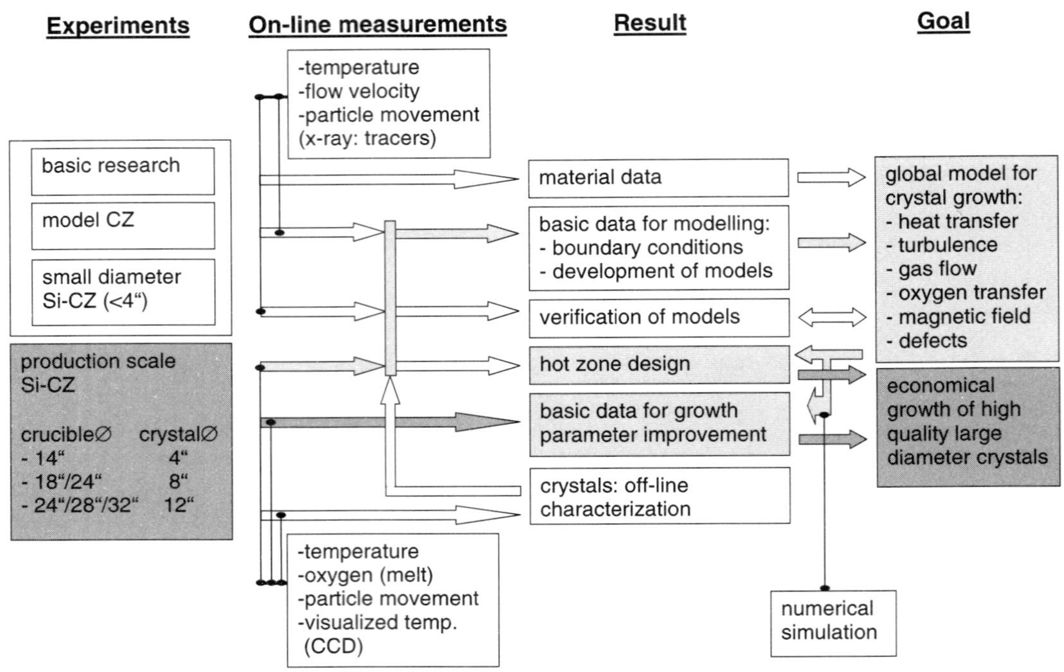

Fig. 2. Schematic diagram of support for process development by experiments and numerical simulation.

Frequently, hot zone modifications were necessary to obtain the required crystal properties. In the case of large diameter process development, the conventional procedure becomes too expensive and time consuming. The support of numerical simulation becomes essential (Fig. 2). Many efforts were made by different groups to experimentally study the physics of crystal growth, e.g.: Material property data as density, thermal expansion coefficient and surface tension of molten Si were measured very precisely [19–21]. In model CZ experiments with small diameter (160 mm) mercury melts, melt convection and the influence of magnetic cusp field were studied [22]. In small diameter (75 mm) silicon melts, the movement of solid tracers was visualized by double-beam X-ray radiography and temperatures measured by thermocouples to investigate the heat transfer [23] and the influence of different cusp field positions [9]. Temperatures distributions were visualized at the surface of large diameter (18″) silicon melts [24].

We preferred to study the melt flow in production scale crucibles (14″–32″), because the melt flow regimes are different in small melts (laminar flow) and large melts (turbulent flow). Experimental set-ups are developed to measure temperature distributions and additionally oxygen concentrations on-line during the growth of crystals up to 300 mm (Fig. 3) [25] (Fig. 4) [26]. By analyzing the time resolved temperature measurements, the convective melt flow was determined in dependence on different parameters. In addition the flow velocity was measured as a function of magnetic field strengths in model CZ InGaSn melts (crucible diameter 14″–20″) [27]. The obtained data are directly used to improve the process parameters and to get the boundary conditions for numerical simulations.

The temperature distributions in the hot zone and crystal are predicted by global thermal numerical models very well [28]. An improvement of the global thermal simulation was obtained by fitting the

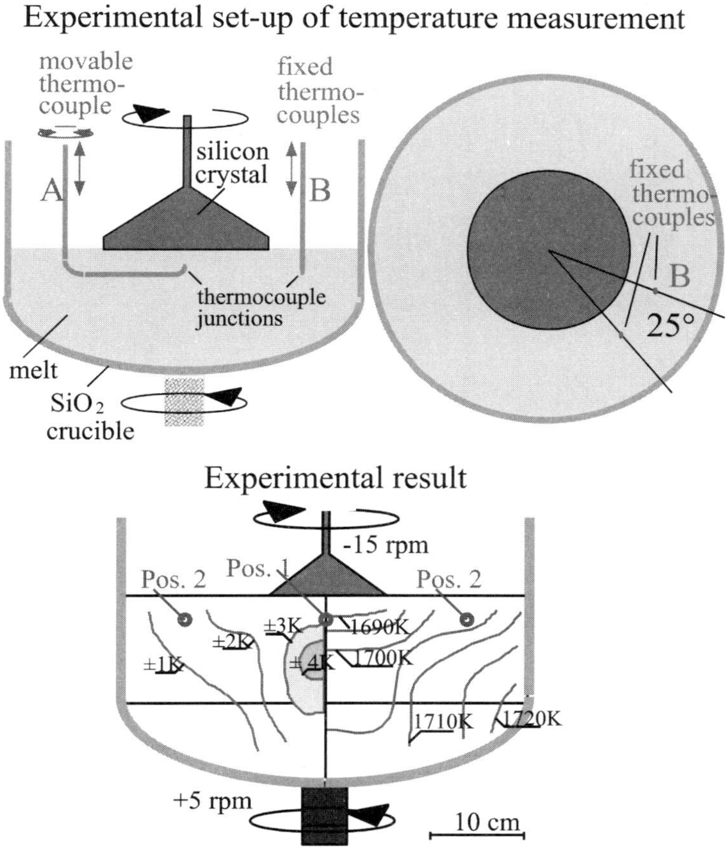

Fig. 3. On-line measurement of temperature in silicon melt during crystal growth and result. a) and b): Schematic drawing of the experimental set-up. One thermocouple (A) can be moved to any position in the melt, two fixed thermocouples (B) are used to measure phase correlations of temperature fluctuations. c) Mean isotherms (right half) and mean maximum amplitudes of temperature fluctuations (left half) measured during growth (200 mm crystal, 70 kg charge weight, 18″ silica crucible) [25].

effective thermal heat coefficient for large melt volumes to the measured temperature distributions [25]. Today we use numerical simulations for all our hot zone design as well as for improving the process conditions to obtain required crystal properties.

However, the actual melt flow is not adequately reproduced in an existing turbulence model. Some progress seems to be obtained in developments of models for the oxygen transport in silicon melts (Fig. 5) [29,30]. A detailed description of the state of the art simulating oxygen transfer is presented in another paper in this volume [31].

2.8. Defect engineering

With respect to the 300 mm crystal growth, one of the most difficult challenges is the achievement of crystal properties, which are required for the 0.18 and 0.13 μm design rules. In Fig. 6 the defect

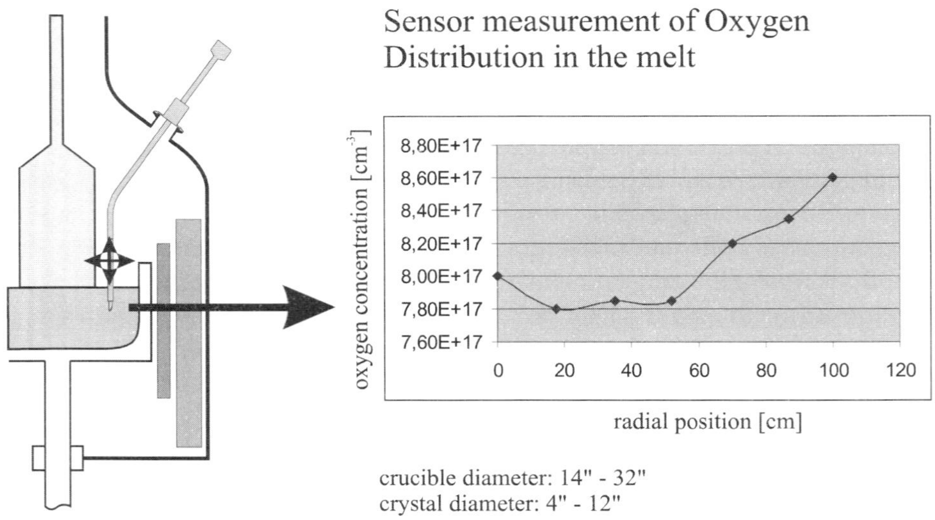

Fig. 4. On-line measurement of oxygen in the melt during crystal growth: a) schematic drawing of experimental set-up, b) measured radial oxygen concentration [26].

distribution of a characterized 300 mm OSF ring wafer is presented. The OSF ring separates the outer interstitial-rich and the inner vacancy-rich region. A good agreement between OSF ring position and GOI yield distribution is observed. It is well known [32–34] that the radial defect distribution strongly depends on the growth conditions. With the empirical formula $v/G(r) = C_{crit} = 1.3 \times 10^{-3}$ cm^2 min^{-1} K^{-1} the radial position of the OSF ring can be calculated for different pull rates in a given hot zone, respectively given axial temperature gradient $G(r)$ (Fig. 7). By lowering the pull rate the diameter of the OSF ring shrinks until it disappears in the center of the crystal on a critical pull rate. However, L-pits, which deteriorate the device yield, are observed. Increasing the pull rate, as done in 200 mm growth, to values such that the OSF ring disappears at the wafer rim can be critical with respect to thermal stress (Fig. 8) and breaking of the ingot becomes probable. The calculated values for the critical stress in Fig. 8 are based on experimental values derived from 300 mm growth experiments. The strategy will be to homogenize v/G radially by hot zone modifications. But it seems to be very difficult to obtain a sufficient improvement, especially with respect to robust growth conditions along the whole crystal length. Otherwise yield problems lead to increasing costs. The wafer annealing, as practiced for 200 mm wafer, may be critical for the 300 mm wafer because of slip generation and would be an additional cost factor. A good alternative is the use of epitaxial wafers if they are tolerated by the device process [35].

2.9. Characterization

In standard bulk characterization two wafers of 150 mm diameter are drilled out from one 300 mm wafer (Fig. 6). The measured data are representative for the whole wafer as the diameter of 150 mm

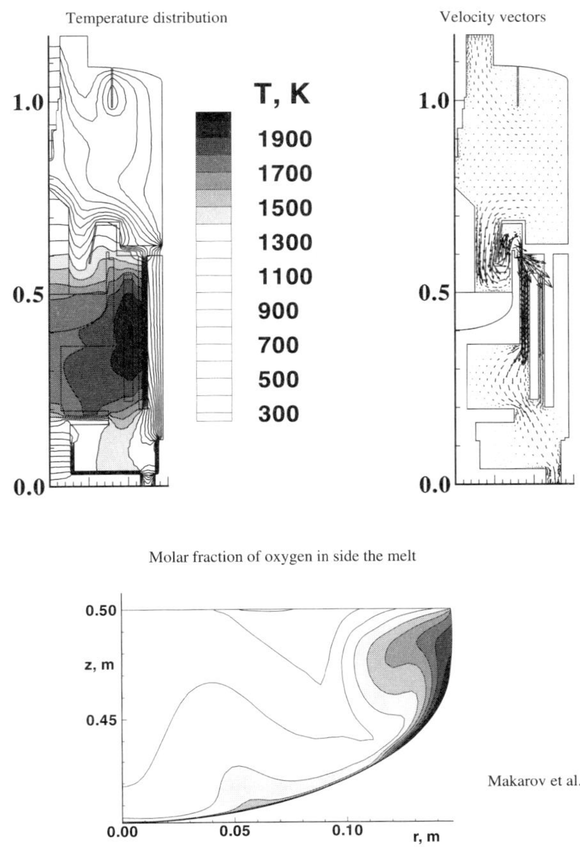

Fig. 5. Numerical simulation of oxygen transport (Air pressure = 10 mbar, flow-rate = 700 LH, crucible rotation = −5 rpm): a) temperature distribution, b) velocity vectors, c) molar fraction of oxygen inside the melt: $x_{min} = 5 \times 10^{-6}$, $x_{max} = 5 \times 10^{-5}$, step $= 5 \times 10^{-6}$ [29].

slices is equal to the radius of the 300 mm wafer. Therefore, material costs can be saved by this method. Two additional 100 mm wafers remain for further investigations.

Additionally, a nondestructive and noncontaminating automatic wafer inspection system provides a one hundred percent measurement, e.g. for oxygen content, to guarantee small tolerances or if necessary, a selection of wafers.

3. Conclusion

The changeover to 300 mm crystal diameter or to 400 or to 450 mm is a large challenge. The development of the key technologies, which are required for the achievement of crystal qualities and economy are started. For the cost development, the crystal manufacturer is not alone responsible, but also the suppliers of consumables and growth equipment and the semiconductor device industry by

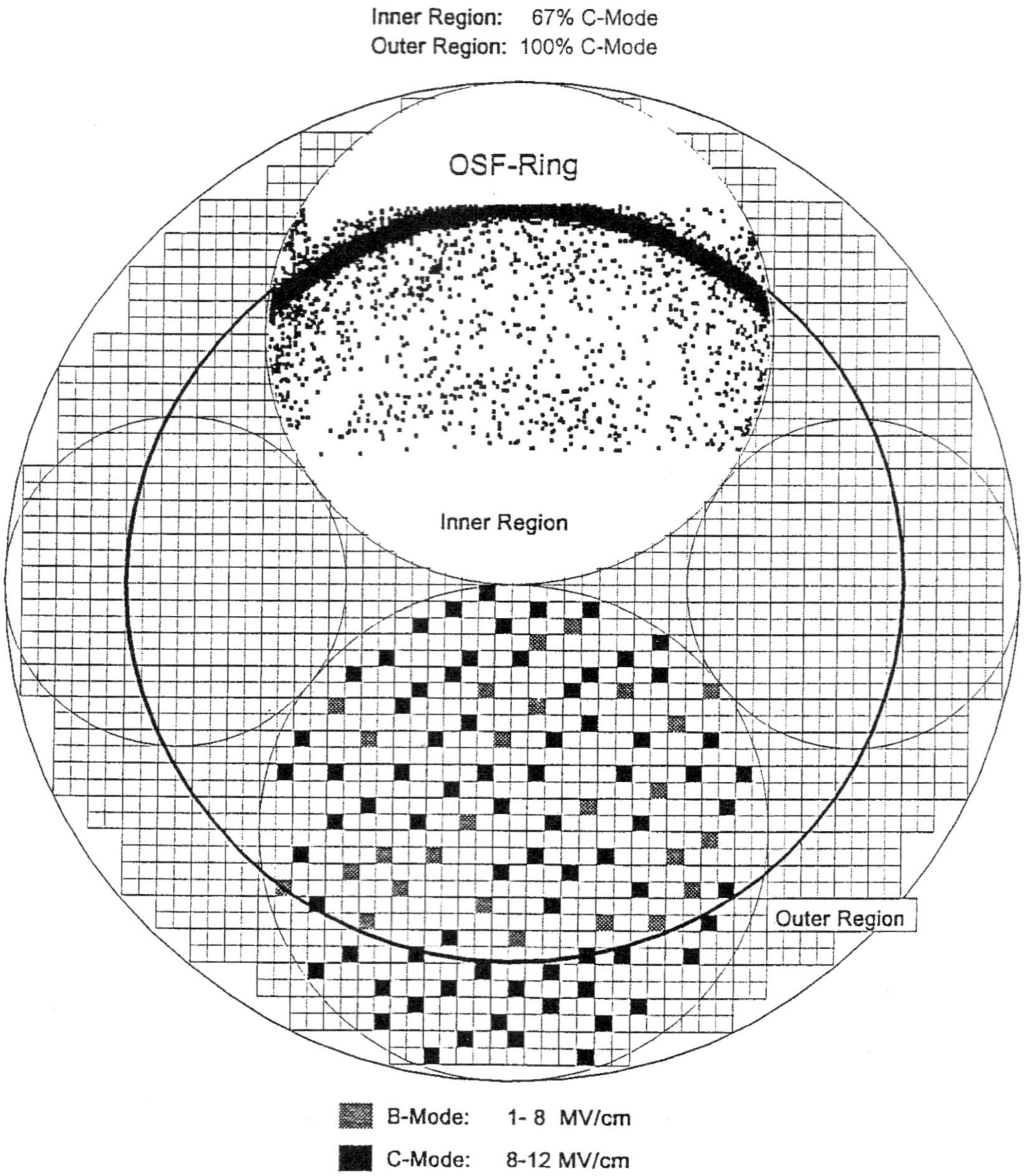

Fig. 6. Characterization results of a 300 mm OSF ring wafer: upper 150 mm wafer: Oxygen induced stacking faults with high density at the ring position after etching (3 min. Secco), lower 150 mm wafer: GOI integrity distribution. Black squares: good capacitors, grey squares: bad capacitors.

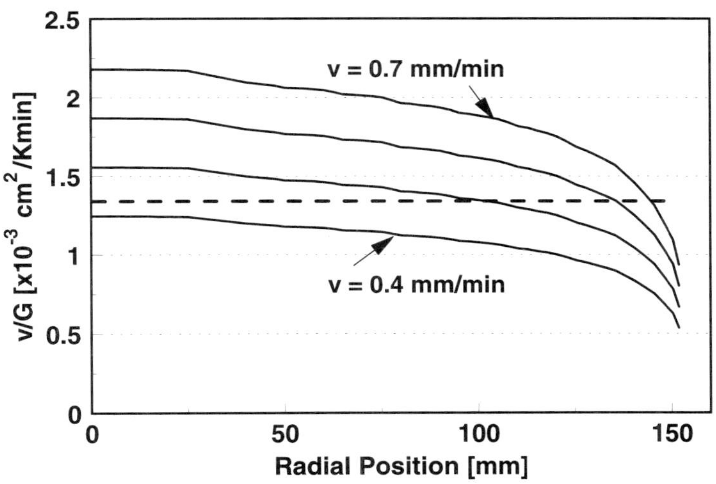

Fig. 7. $v/G(r)$ calculated for a 300 mm crystal grown with a given hot zone versus the radial position for different pull rates v with a spacing of 0.1 mm/min. The dotted line represents C_{crit}.

specifying the crystal qualities. The transition to the epi wafer offers an interesting alternative, because the required properties can be better obtained in many cases as compared to the polished wafer and the price ratio between epi and polished wafer will decrease. Overall, the changeover to larger diameters will provide a higher productivity and capacity for the device industry [1].

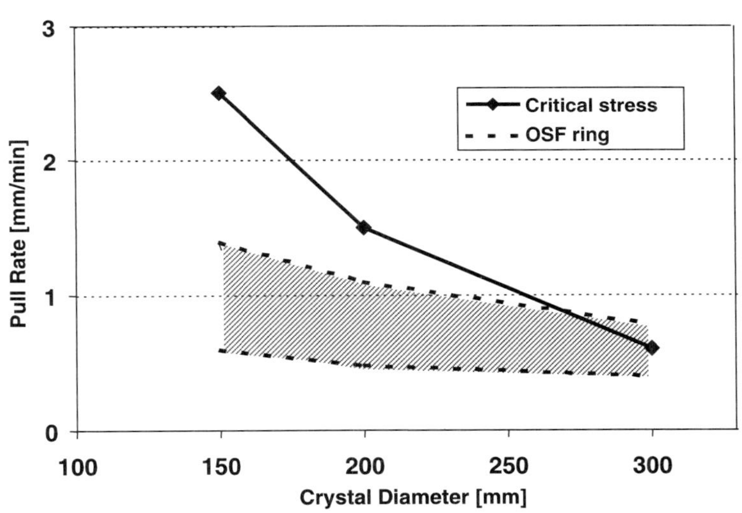

Fig. 8. (♦) denotes the maximum pull rate limited by the critical thermal stress (calculated for a given hot zone type) at which the crystal is likely to crack during growth. The shaded area shows the pull rate range within which an OSF ring appears on the wafer.

Acknowledgements

The authors would like to thank Prof. Müller, Dr. Seidl, Mr. Graebner, Mr. Mühe and Mr. Axmann for developing and perfoming the on-line measurements and Prof. Gelfgat and Dr. Gorbunov for the CZ model experiments. This work has been supported by the Federal Department of Education, Science, Research and Technology of FRG under contract number 01M2973A. The authors alone are responsible for the contents of this publication.

References

[1] H.R. Huff, R.K. Goodall, Silicon Wafer Thermal Processing 300 mm Issues, http://www.semiconductor.net/semiconductor/archive/May98/reference/ff3.html.
[2] D. Seligson, The economics of 300 mm processing, Semiconductor International, 1998, pp. 52–58.
[3] K. Takada, H. Yamagishi, H. Minami, M. Imai, Research and development of super silicon wafer, Electrochem. Soc. Proc. 98-1 (1998) 376–395.
[4] Private communication by Leybold.
[5] H. Yamagishi, M. Kuramoto, Y. Shiraishi, M. Machida, K. Takano, N. Takase, T. Iida, J. Matubara, H. Minami, M. Imai, K. Takada, Large diameter silicon technology and epitaxy, in: E-MRS Proc. F, Straßburg, Spring 1998.
[6] Private communication by Kayex.
[7] T. Kanda, M. Hourai, S. Miki, T. Shigematsu, H. Tomokage, T. Miyano, H. Morita, A. Shintani, Influence of melt-temperature fluctuations on striation formation in large-scale Czochralski Si growth systems, J. Cryst. Growth 166 (1996) 663–668.
[8] H. Hirata, K. Hoshikawa, N. Inoue, Improvement of thermal symmetry in CZ silicon melts by the application of a vertical magnetic field, J. Cryst. Growth 70 (1984) 330–334.
[9] M. Watanabe, The effect of cusp configuration on temperature and flow fields in a MCZ melt, Electrochem. Soc. 98-1 (1998) 441–452.
[10] K. Takada, The ultimate challenge for the 21st century: prospect of ∅400 mm silicon wafers and technological break-throughs, in: Proc. of the Kazusa Akademia Park Forum on the Science and Technology of Silicon Materials, 1997, pp. 15–20.
[11] H.-D. Chiou, T.Y.T. Lee, S. Teng, The stress and strength at the neck of a large diameter silicon crystal during growth, J. Electrochem. Soc. 144 (8) (1997) 2881–2886.
[12] H. Yamagishi, H. Minami, M. Imai, T. Misawa, K. Takada, CZ crystal growth development in super silicon crystal project, in: Proc. 2nd Int. Symp. Adv. Sci. Techn. Silicon Mat., Kona, 1996, pp. 59–63.
[13] K.M. Kim, P. Smetana, Maximum length of large diameter Czochralski silicon single crystals at fracture stress limit of seed, J. Cryst. Growth 100 (1990) 527–528.
[14] A. Takao, A history and future of silicon crystal growth, Electrochem. Soc. Proc. 98-1 (1998) 157–179.
[15] S. Chandrasekhar, K.M. Kim, Growth of large diameter (>10 mm) necks for large size CZ silicon, Electrochem. Soc. Proc. 98-1 (1998) 411–416.
[16] T. Abe, Innovated silicon crystal growth and wafering technologies, in: ULSI Science and Technology, Proceedings, Vol. 97-3, The Electrochemical Society, Pennington, NY, 1997, p. 123.
[17] R.L. Hansen, R.D. Shelly, L.E. Drafall, R.M. McCutchan, J.D. Holder, L.A. Allan, European Patent Application, EP 0 753 605 A1.
[18] Private communications by different silicon crucible suppliers.
[19] K. Terashima, H. Nakanishi, S. Maeda, K. Abe, K. Hoshikawa, S. Kimura, New approach to CZ Si growth, in: Proc. 2nd Int. Symp. Adv. Sci. Techn. Silicon Mat., Kona, 1996, pp. 64–71.
[20] M. Langen, T. Hibiya, M. Eguchi, I. Egry, Measurement of the density and the thermal expansion coefficient of molten silicon using electromagnetic levitation, J. Cryst. Growth (1998), in press.
[21] M. Przyborowski, T. Hibiya, M. Eguchi, I. Egry, Surface tension measurement of molten silicon by the oscillating drop method using electromagnetic levitation, J. Cryst. Growth 151 (1995) 60–65.

[22] Y.S. Lee, Ch.-H. Chun, Experiments on the oscillatory of low Prandtl number liquid in Czochralski configuration for crystal growth with cusp magnetic field, J. Cryst. Growth 180 (1997) 477–486.
[23] M. Watanabe, K. Kakimoto, M. Eguchi, T. Hibiya, Modification of heat and mass transfers and their effect on the crystal-melt interface shape of Si single crystal during Czochralski crystal growth, Jpn. J. Appl. Phys. 36 (1997) 6181–6186.
[24] M. Tanaka, M. Hasebe, N. Saito, Pattern transition of temperature distribution at Czochralski silicon melt surface, J. Cryst. Growth 180 (1997) 487–496.
[25] A. Seidl, G. Müller, E. Dornberger, E. Tomzig, W.v. Ammon, Turbulent melt convection and its implication on large diameter silicon Czochralski crystal growth, Electrochem. Soc. 98-1 (1998) 417–428.
[26] A. Seidl, R. Marten, G. Müller, In situ investigation of oxygen distribution and transport in Czochralski silicon melts by electrochemical solid ionic sensors, J. Cryst. Growth 166 (1996) 680–684.
[27] Y. Gelfgat, L. Garbunov, W.v. Ammon, J. Virbulis, E. Tomzig, to be published.
[28] E. Dornberger, E. Tomzig, A. Seidl, S. Schmitt, H.-J. Leister, Ch. Schmitt, G. Müller, Thermal simulation of the Czochralski silicon growth process by three different models and comparison with experimental results, J. Cryst. Growth 180 (1997) 461–467.
[29] Yu.E. Egorov, Yu.N. Makarov, E.A. Rudinsky, E.M. Smirnov, A.I. Zhmakin, Modelling analysis of oxygen transport during Czochralski growth of silicon crystals, in: MRS Proc., Fall 1998, Boston, in press.
[30] A. Mühe, R. Backofen, G. Müller, E. Tomzig, W.v. Ammon, Oxygen distribution in silicon melt during a standard Czochralski process studied by sensor measurements and comparison to numerical simulation, presented ICCG, Jerusalem, 1998.
[31] G. Müller, R. Backofen, A. Mühe, W.v. Ammon, E. Tomzig, Study of oxygen transport in Czochralski growth of silicon, in: EMRS Proc. F./Straßbourg, 1998.
[32] W.v. Ammon, E. Dornberger, H. Oelkrug, H. Weidner, The dependence of bulk defects on the axial temperature gradient of silicon crystals during Czochralski growth, J. Cryst. Growth 151 (1995) 273.
[33] E. Dornberger, W.v. Ammon, The dependence of ring-like distributed stacking faults on the axial temperature grandient of growing Czochralski silicon crystals, in: J. Electrochem. Soc. 143 (1996) and Proceedings of the ECS Spring Meeting, 1995.
[34] E. Dornberger, W.v. Ammon, N. Van den Bogaert, F. Dupret, Transient computer simulation of a CZ-crystal growth process, J. Cryst. Growth 171 (1996) 452.
[35] W.v. Ammon, E. Dornberger, P.O. Hanson, Bulk properties of very large diameter silicon single crystals, presented ICCG, Jerusalem, 1998.

ELSEVIER

Microelectronic Engineering 45 (1999) 127–133

MICROELECTRONIC ENGINEERING

300 mm Epitaxy: challenges and opportunities from a wafer manufacturer's point of view

Per-Ove Hansson*, Martin Fuerfanger

Wacker Siltronic AG, P.O. Box 1140, D-84479 Burghausen, Germany

Abstract

Results on 300 mm silicon wafer epitaxy according to the 0.18 μm design rule requirements are presented. Wafer uniformity, surface metals, geometry, and thermal stability demonstrate production capabilities. Localized light scattering particles and crystal defects are identified as the largest technological challenge. Preliminary experiments indicate that industry's requirements can be met. The p/p + epitaxial wafer is the best candidate to meet the 300 mm cost challenge, as well as the technological requirements for integrated circuit manufacturing.

1. Introduction

The major keys to achieve higher IC functionality are smaller lithography design rules (DR) and larger chip areas. Simultaneously the market demands lower cost per logic function, which have led to a cost per bit reduction of nearly 30% per year the last 30 years [1]. Therefore, the semiconductor industry must continue to reduce manufacturing cost. A significant contribution to cost improvement is the transition to larger wafers, since the number of transistors per chip continues to grow faster than the chip area can be decreased by smaller DR and chip layout [2]. The semiconductor industry is currently preparing the step to 300 mm diameter wafers. The 300 mm wafers, as compared with 200 mm wafers, provide 25–30% lower IC cost per die, the usable fraction of the wafer increases up to 14% due to the smaller wafer curvature, and the equipment capital cost increases 20–40%. These changes in wafer economics lead to a lowered cost per chip of 27–39%, depending on the device type [3]. Several IC manufacturers have targeted 300 mm pilot lines for 1999. Leading wafer suppliers have already installed 300 mm pilot production lines.

The 300 mm conversion involves technological challenges both relating to the increase in wafer diameter and to the simultaneous shift in DR to 0.18 μm. The Semiconductor Industry Association (SIA) National Technology Roadmap for Semiconductors outlines the semiconductor industry's requirements for the new generations of Si wafers. Table 1 shows selected parameters and timelines.

*Corresponding author.

0167-9317/99/$ – see front matter
PII: S0167-9317(99)00110-0

Table 1
SIA National Technology Roadmap for Semiconductors 1997, starting materials technology requirements: selected data

SIA Semiconductor Roadmap data	1997	1999	2001
General characteristics			
Technology generation	0.25 μm	0.18 μm	0.15 μm
Wafer diameter (mm)	200	300	300
Site flatness (SFQR) (nm)	250	180	150
Site size (mm^2)	22 × 22	25 × 32	25 × 34
Localized light scatterers (LLS), incl. particles (per wafer)	172	192	178
LLS size (nm)	125	90	75
Particles (per wafer)	41	60	47
Epitaxial wafer			
Layer thickness (μm) (± %tolerance)	2–5 (±5%)	2–4 (±4%)	2–4 (±4%)
Layer structural defects (per wafer)	1	2	1.8

These technological requirements, combined with the increases in manufacturing cost, mainly caused by a lowered ingot yield and capital cost for new technologies, make the commercial challenge of the 300 mm conversion highly significant for wafer manufacturers.

Epitaxial wafers are a well established choice for advanced integrated circuits like microprocessors and other high value logic devices. The epitaxial wafer offers a number of advantages as compared to the polished and polished annealed wafer: the surface is virtually free of crystal originated particles (COP) and other defects. Gate oxide integrity (GOI) values are, therefore, considerably better [4]. Latch up immunity of devices is achieved when p+ substrates are used [7]. Metallic gettering of e.g. Fe due to the formation of Fe–B pairs depletes the epitaxial layer of Fe atoms by trapping them in the substrate [5,6].

This paper discusses 300 mm epitaxy technology issues like thermal stress, epitaxial layer uniformity, and epitaxy crystal defect generation in the light of the industry's new requirements. It is shown that the 300 mm epitaxial wafer meets the demands of the semiconductor industry.

2. Results of 300 mm epitaxy process development

A series of experiments was performed to develop and characterize 300 mm epitaxy processes for 0.18 μm design rule requirements. The substrates were produced by Wacker and are p− and p+ furnace- and particle grade wafers. The p+ wafers have a low temperature oxide (LTO) backseal. The epitaxy processes were run on an Applied Materials 300 mm Epi Centura™ tool at atmospheric pressure with trichlorosilane (TCS) as source gas. Deposition temperatures were chosen around 1100°C. The experiments were performed in a laboratory environment rather than in a mass production facility. Results are compared with the current International 300 mm Initiative (I300I) specification for 0.18 μm DR wafers [8].

2.1. Thermal stability and slip line generation

The higher the ramp rates and the higher the temperatures the more thermal stress is generated in the wafer, which causes a higher probability of plastic deformation, so called slip line (or dislocation), generation [9]. In 300 mm thermal processing the stress induced by gravitation, which is proportional to the square of the wafer diameter and inversely proportional to the wafer thickness, is estimated to influence the slip line generation, since the ratio of diameter to thickness of 300 mm wafers increases to 387, compared to 276 for 200 mm wafers.

The thermal stability during the epitaxy process was studied by means of applying different temperatures, thermal gradients, and temperature ramps to the 300 mm Si substrate during epitaxy processing in the Epi Centura™ tool. After the thermal processing the wafers were investigated by Normaski interference differential contrast (NDIC) microscopy without edge exclusion. Bakes at temperatures between 1100°C and 1200°C for 60 s combined with Si deposition at temperatures between 1100°C and 1140°C for 60 s. The maximum temperatures without any slip line generation were 1200°C for the bake and 1140°C for the deposition. After the definition of the maximum temperatures, the temperature ramp rate from 700°C upwards was maximized. The maximum ramp rate without slip generation was 10°C/s. Our experiments show that thermal processing in an epitaxy reactor is definitely possible, under the right process conditions.

2.2. Uniformity of layer thickness and resistivity

Deposition with TCS as source gas generally takes place in the temperature transition region between diffusion controlled and surface kinetic controlled growth [10]. Therefore, the key to uniformity is controlling both gas flows and temperature uniformity in the deposition chamber. Fig. 1 shows contour plots of layer thickness and resistivity of a 300 mm p/p + (LTO) wafer. The relative ranges (max − min/average) of thickness and resistivity within wafer are after process tuning in general below ±1.5% and ±3%, respectively, within one cassette (13 wafers), as measured with a 17 point 'star' or with a radial line pattern. The peripheral points are positioned 3 mm from the edge.

Fig. 2 shows the variation of thickness and resistivity averages over 500 p/p + (LTO) wafers, without process tuning. Every 13th wafer was evaluated. A slight drift of the mean values due to the

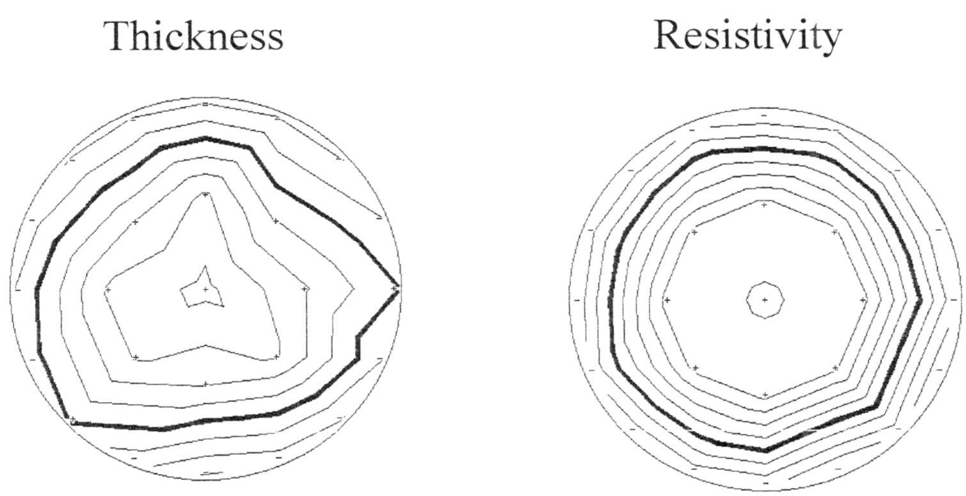

Fig. 1. Thickness (left) and resistivity (right) contour plots of a 300 mm epitaxy wafer. Thickness: mean value (thick line) 3.15 μm, rel. range (max − min/average) ±1.1%, contour line distance 0.01 μm. Resistivity: mean value 10.60 Ω cm, rel. range ±2.1%, contour line distance 0.05 Ω cm.

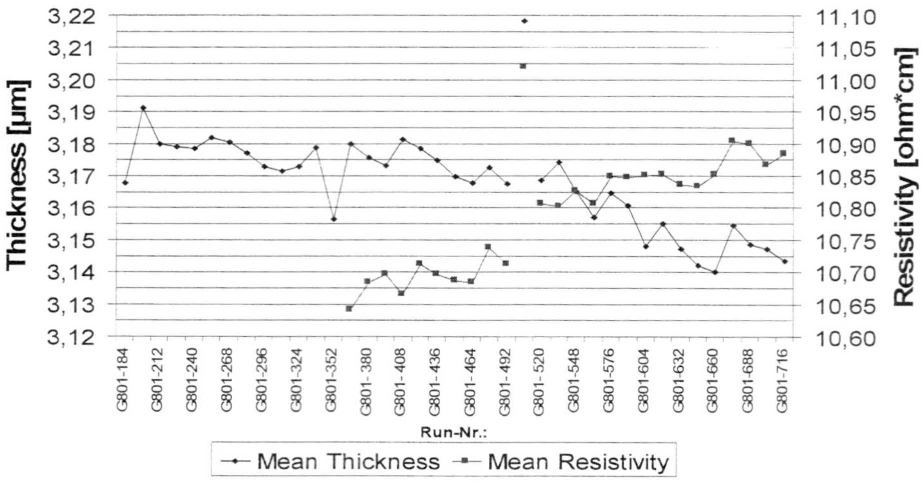

Fig. 2. Variation of thickness and resistivity mean values of 500 p/p+(LTO) wafers, without process tuning. Every 13th wafer was measured. The relative range of thickness and resistivity averages is ±1.2% and ±1.8%, respectively.

lack of statistical process control (SPC) and production proof gas facilities during testing can be observed in Fig. 2. The relative range of the within wafer averages was 1.2% for thickness and 1.8% for resistivity. The relative range of all 9044 measured values was ±2.4% and ±3.6% (2σ) for thickness and resistivity, respectively. The I300I specification defines a thickness variation of ±4% and a resistivity variation of ±5% within one wafer. Therefore, the produced values indeed show mass production capability of 0.18 µm DR wafers.

2.3. Localized light scatterers and epitaxy crystal defects

The high localized light scatterer (LLS) and crystal defect requirements (see Table 1) for the 0.18 µm design rule may be the most significant technological challenge. During Czochralski crystal growth the ratio of the pull rate to temperature at the growth interface may not be kept over the critical value for bulk Si interstitials formation over the whole crystal volume. These interstitials may lead to the formation of dislocation loops, so called L-pits, at the edge of the low doping level ($<10^{16}$ cm^{-3}) p− crystals [11]. These L-pits may cause severe damage to devices due to their extension of several µm. Therefore, compared with 200 mm crystals, larger volumes of crystal are scrapped, which adds to the crystal manufacturing cost together with the lower pull rate [12]. Recent studies show that the L-pits generated by the 300 mm p− crystal pulling process create defects in the epitaxial layer [11]. Therefore, epitaxy may not be able to 'repair' p− substrates originating from the part of the crystal which contains L-pits. However, p/p+ wafers containing a substrate boron concentration $\geq 10^{19}$ cm^{-3} not only suppresses L-pit formation [11] but also reduce the COP content to virtually zero [4]. Therefore, the crystal pulling manufacturing cost for 300 mm p/p+ epitaxial wafers will be lower, due to the higher crystal pulling yield, as compared to p− polished wafers, which increases the incentive for using 300 mm epitaxial p/p+ wafers in IC manufacturing.

In Fig. 3 LLS values of 50 p/p+(LTO) wafers after epitaxy and final cleaning are shown in a percentile plot. The measurements were obtained by Tencor SP1™, using the dark field wide (DFW)

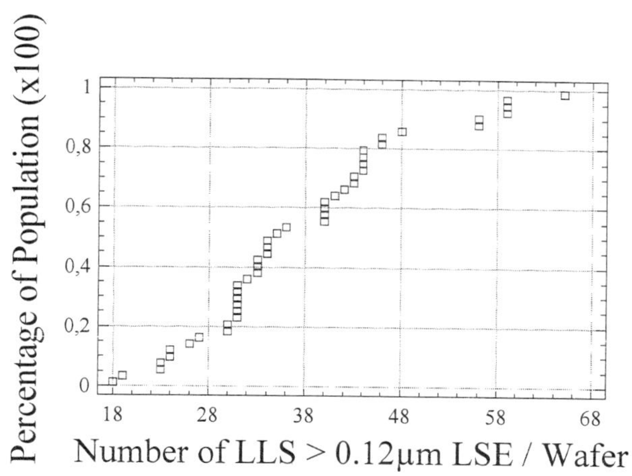

Fig. 3. LLS values of 50 p/p + (LTO) wafers, manufactured under non-sufficient clean room conditions, after epitaxy and final cleaning. The percentage of the wafer population with a certain maximum LLS value is plotted against the LLS per wafer value. Tencor SP1, LSE >0.12 μm LSE, DFW.

channel at a latex sphere equivalent (LSE) of ≥0.12 μm. A chosen combination of different measurement channels of the Tencor SP1™ allows the discrimination of COP, particles, and epitaxy crystal defects [13]. The p/p+ epitaxy wafers show no COP counts. The LLS 2σ value is 59 per wafer, mainly caused by particles on the wafer surface. The I300I specification for 0.18 μm DR allows 100 LLS at ≥0.12 μm LSE for furnace wafers and 100 LLS at ≥0.09 μm LSE for particle wafers. The LLS values, which meet the furnace wafer specification, were obtained under laboratory conditions with non-sufficient clean room capability. Thus a potential for meeting the particle wafer specification under production conditions is given. Data from wafers manufactured in a 300 mm production environment, where lower LLS values are expected, will be available at the end of 1998.

2.4. Surface metals

Fig. 4 shows the metal concentrations on the surface of five randomly chosen wafers directly after epitaxy, as measured by inductively coupled plasma mass spectroscopy (ICPMS) and total reflection X-ray fluorescence (TXRF). All measured metal concentrations except for S, Fe, Zn, and Br are

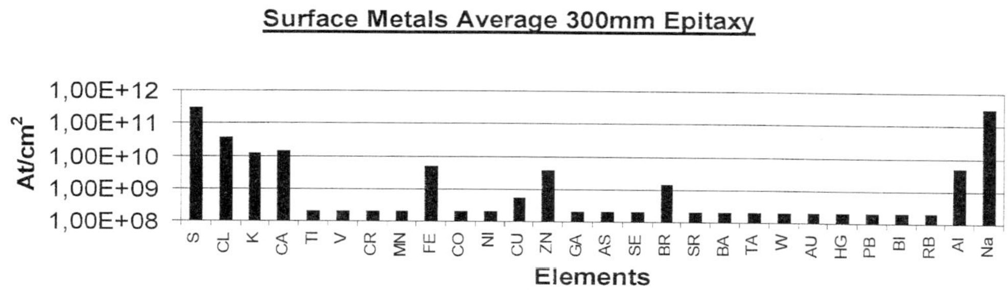

Fig. 4. Surface metal values of five randomly chosen wafers after epitaxy. The bars represent the detection limit, except for S, Fe, Zn, and Br, where values are slightly above.

below the detection limits of ICPMS and TXRF. All values, including Fe, Zn, and Br, are below the current I300I specification limits for the 0.18 μm DR: 10^{10} cm^{-2}. Sulfur, which is not specified, is below 10^{12} cm^{-2}. Specifications for organic contamination are currently being established.

2.5. Geometry and wafer backside

Geometry, i.e. flatness, data of the 300 mm epitaxy wafers do not significantly deviate from the substrate values. Warp values of p+ wafers with an LTO backseal of 250 nm and a 3 μm thick 10 Ω cm epitaxial layer are in the 50 μm range (the current I300I 0.18 μm DR specification is <100 μm for furnace wafers, <50 μm for lithography wafers). These relatively high values can be ascribed to the differences in thermal expansion coefficients of the layers and the substrate, leading to a large scale elastic deformation of the wafer upon cool down to room temperature after the deposition processes. Thinner LTO layers lower warp (or bow) values in general [14].

The polished backside of the 300 mm wafer is a major difference to the 200 mm standard. Marks originating from the contact to the susceptor during epitaxy are more visible as compared to 200 mm wafers, which have etched or other non-polished backsides. These backside marks are more of a cosmetic issue than a factor influencing device yield. A specification for 'harmless' backside defects is included in the I300I specification for 0.18 μm DR [8].

3. Conclusions

Epitaxy of 300 mm Si wafers according to 0.18 μm DR specifications and roadmaps is indeed possible. Results on wafer uniformity, geometry, surface metals, and thermal stability show production capabilities. Further evaluation of LLS capability, which is identified as the largest technological challenge, in a mass production environment is necessary, however, preliminary laboratory experiments indicate that the industry's requirements can be met. The commercial challenge of meeting the cost targets for 300 mm epitaxial wafers is superior to the technological issues. The best candidate to meet the cost requirements is the p/p+ epitaxial wafer, due to the higher crystal pulling yield as compared to p/p− wafers. The p/p+ wafer also offers several technological advantages, i.e. a COP surface density of virtually zero, internal gettering, higher GOI yield, and latch up immunity.

Acknowledgements

The work has been supported by the Federal Department of Education, Science, Research, and Technology of the FRG under contract number 01 M 2973 A. The authors alone are responsible for the contents of this publication.

References

[1] G.E. Moore, Memories and the microprocessor, Daedelus 125 (2) (1996) 55–80.
[2] H. Huff, R. Goodall, Solid State Phenomena 47–48 (1996) 65.
[3] M. Telford, Eur. Semicond. 14 (1997) 13–14.

[4] D. Gräf, M. Suhren, U. Lambert, R. Schmolke, A. Ehlert, W. von Ammon, P. Wagner, J. Electrochem. Soc. 145 (1) (1998) 275–284.
[5] D. Gilles, W. Schroeter, W. Bergholz, Phys. Rev. B 41 (1980) 5770.
[6] M. Aoki, T. Itakura, N. Sasaki, Appl. Phys. Lett. 66 (20) (1995) 2709–2711.
[7] R.B. Herring, in: W.C. O'Mara, R.B. Herring, L.P. Hunt (Eds.), Handbook of Semiconductor Silicon Technology, Noyes Publications, New Jersey, USA, 1990, pp. 248–336.
[8] See, e.g. the I300I Internet Homepage, http://www.i300i.org
[9] H.R. Huff, R.K. Goodall, in: R.B. Fair, B. Lojek (Eds.), Proceedings of the 3rd International Rapid Thermal Processing Conference, RTP 95, 1995, pp. 9–40.
[10] H.M. Liaw, J.W. Rose, in: B.J. Baliga (Ed.), Epitaxial Silicon Technology, Academic Press, Orlando, 1986, pp. 1–85.
[11] W. v. Ammon, E. Dornberger, P. O. Hansson, Proceedings of ICCG 12, 1998, J. Cryst. Growth, to be published.
[12] E. Tomzig, W. v. Ammon, E. Dornberger, U. Lambert, W. Zulehner, Proceedings of the E-MRS Spring Meeting 1998, Symposium F, these proceedings.
[13] F. Passek, H. Piontek, A. Luger, P. Wagner, Proceedings of the E-MRS Spring Meeting 1998, Symposium F, these proceedings.
[14] E.W. Hearn, D.J. Werner, D.A. Doney, Solid State Sci. Technol. 133 (8) (1986) 1749–1751.

ELSEVIER

Microelectronic Engineering 1 (1999) 135–147

MICROELECTRONIC ENGINEERING

Study of oxygen transport in Czochralski growth of silicon

G. Müller[a], A. Mühe[a,*], R. Backofen[a], E. Tomzig[b], W.v. Ammon[b]

[a]*Department of Materials Science WW 6, Crystal Laboratory of the University Erlangen-Nürnberg and Fraunhofer Institute IIS-B, Martensstr. 7, D-91058 Erlangen, Germany*

[b]*Wacker Siltronic AG, P.O. Box 1140, D-84479 Burghausen, Germany*

Abstract

The present status and special features of experimental analysis and numerical modelling of oxygen transport in the Czochralski growth of silicon is treated. The solubility of oxygen in liquid Si is discussed in detail as this parameter is used as a boundary condition (oxygen source) in the numerical process models. The possibility of in situ measurement of the oxygen distribution in the melt in dependence from boundary conditions by using an electrochemical sensor is shown.

Model conceptions for a numerical simulation of the oxygen transport in Czochralski melts are discussed in detail. The problem of boundary layer resolution is analysed, examples how to solve it are given. A quantitative modelling of the oxygen transport suffers at present from the fact that the existing turbulence models are not adequate to predict the velocity field precisely enough.

Keywords: Silicon; Czochralski crystal growth; Oxygen measurement; Numerical simulation; Transport modelling

1. Introduction

Toady's processes for the fabrication of IC's are based on the so-called intrinsic gettering process which requires a precisely defined concentration of oxygen (about $5*10^{-17}$ cm^{-3}) with uniform distribution in the Si wafers. Part of this oxygen is reacting with point defects during thermal treatment to form precipitates surrounded by other crystallographic defects which are acting as these gettering centers. Part of the oxygen remains solved as interstitial point defect which is necessary to harden the Si lattice. The oxygen is incorporated into the Si crystal during the Czochralski growth process. The source of oxygen is the chemical reaction of the Si melt with the SiO_2 crucible (see Fig. 1).

The oxygen concentration incorporated into the crystal is determined by the oxygen concentration in the melt at the growth interface. However, the oxygen concentration found in the crystals as well as in quenched portions of the melt is always significantly lower than the solubility because a high percentage of dissolved oxygen evaporates as SiO from the free surface of the melt. This indicates

*Corresponding author. Tel.: +49-9131-761-332; fax: +49-9131-761-390.
E-mail address: a.muehe@ww.uni-erlangen.de (A. Mühe)

0167-9317/99/$ – see front matter
PII: S0167-9317(99)00115-X

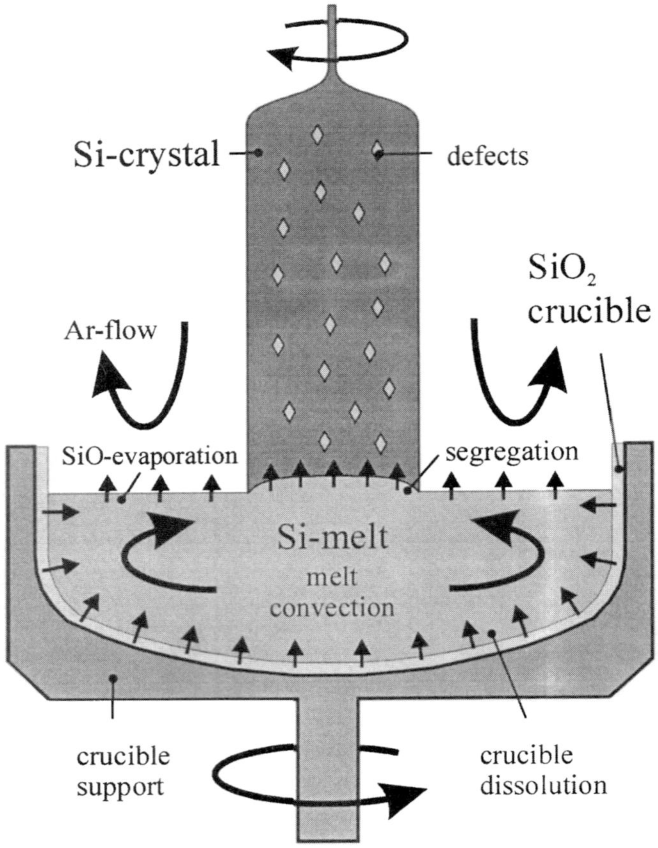

Fig. 1. Schematic representation of relevant processes of the oxygen transport during the Czochralski growth of silicon single crystals.

that there is typically no equilibrium state between the SiO_2 crucible wall and the bulk silicon melt. Estimations of the total amount of oxygen dissolved into the melt per unit time by measurement of the SiO_2 dissolution rate under Czochralski conditions and comparison to the amount of oxygen found in the crystal exhibit that only about 1 percent of the oxygen transported through the melt is incorporated in the crystal. This means that the melt oxygen concentration is practically determined by the dynamical equilibrium of the oxygen fluxes into the melt by SiO_2 dissolution (source) and out of the melt by SiO evaporation (drain). Both fluxes depend strongly on the local temperature and on concentration gradients. The portion of oxygen which is incorporated into the crystal and in which we are most interested is negligible with respect to the dynamical equilibrium.

Considering the CZ-configuration as sketched in Fig. 1 it is obvious that the oxygen transport is strongly influenced by the ratio of the crucible surface in contact with the melt to the melt free surface (i), the melt flow structure (ii), and the Ar-flow in the vicinity of the melt surface eventually combined with the surface capillary flow (iii).

In fact, a continuous decrease of oxygen concentration in the crystals is usually found along the growth axis which is attributed to (i). However, correlations with the melt flow and its structure (ii) are not as clear, although we must assume that its influence on the oxygen concentration at the crystal–melt interface is of the same importance as compared to (i) and (iii). For instance it determines whether oxygen-rich melt from the crucible bottom or melt of lower oxygen concentration from the free surface is transported to the growth interface. Furthermore, it influences the heat transport through the melt volume and, therefore, the temperature distributions along the melt–crucible interface and the free melt surface. With respect to the radial distribution of oxygen in the crystals, the crystal rotation is found to be of major influence. While the outward directed rotating-disc flow beneath the crystal causes the oxygen concentration to be distributed homogenuously, the inward directed flow beneath the free melt surface driven by buoyancy and Marangoni convection perturbates the uniformity close to the crystals edge.

In this paper we try to critically review the present state of knowledge coming from the experimental analysis of the oxygen transport in the Si melts under CZ conditions. We also analyse the task of a numerical modelling of the oxygen transport in the CZ process and its present status.

2. Experimental analysis of oxygen transport in silicon melts under Czochralski conditions

2.1. Principles of oxygen detection in silicon melts

Until recently, the information that was available on the dependency of the oxygen transport on growth conditions during the Czochralski process came only from the analysis of the crystals by FTIR and other techniques. As mentioned above, these dependencies are very complex. Therefore, a large quantity of growth-and-analysis circles would be necessary in order to understand and optimize the effects of the various parameters during the Czochralski process. Since increasing crystal diameters require larger furnaces and crucibles with increasing complexity of the melt flow such optimizations have to be carried out over and over again, while the costs for each experiment also keep increasing.

To overcome these problems the use of electrochemical potential sensors was developed [1] to measure the melt oxygen concentration in-situ during the CZ crystal growth process conditions.

2.1.1. Determination of the oxygen solubility in silicon melt

According to the situation illustrated in Fig. 1 we have stated that the dissolving SiO_2 crucible wall is the oxygen source. The most common boundary condition for the melt–crucible interface is to assume that the oxygen concentration in the direct vicinity of the interface equals the solubility concentration at the present temperature. Therefore, a great interest exists in the experimental determination of equilibrium concentrations of oxygen in the $Si|SiO_2$ system. It will also be seen, that the calibration of electrochemical oxygen sensors for the in-situ measurement of the melt oxygen concentration is closely related to the determination of the free enthalpy of solution of oxygen in Silicon melt.

In the literature several attempts of a determination of the oxygen solubility in liquid Silicon were reported. Generally, all methods start with an equilibration procedure of the silicon melt with SiO_2 and/or O_2 and/or SiO at a certain temperature during a certain time. After that, the melt is quenched with a certain cooling rate and finally the solidified sample is analysed by different analysing techniques.

Ekkhult and Carlberg [2] filled silica ampoules with several grams of highly pure polysilicon and sealed them after evacuation. Then they kept the specimen at temperatures between 1415°C and 1490°C for 45 minutes. After quenching they analysed the oxygen concentration in the specimen by secondary ion mass spectroscopy (SIMS). Their results are denoted by triangles in Fig. 2. No significant dependency of the solubility on the temperature was found. The authors compared the experimental data with their results of thermodynamic calculations for equilibria with SiO_2 and amorphous SiO, the former of which is also denoted in Fig. 2 by a solid line. While these calculations did exhibit a temperature dependency, the experimental data lay between the theoretical curves, which the authors interpreted with the possible presence of both SiO_2 and amorphous SiO.

Hirata and Hoshikawa [3] melted a 13 mm diameter CZ-silicon rod in an evacuated silica ampoule in order to make sure a complete enclosure of the silicon by SiO_2. They kept the specimen at temperatures between 1698 and 1820 K for 90 minutes each. Then they quenched the specimen. From the center of the quenched specimen they took a small sample and analysed it by inert gas fusion (IGF) analysis. This technique can be calibrated by using Czochralski-grown silicon single crystals with an oxygen concentration known from FTIR measurements. However, in this case the analysis is not independent from the FTIR analysis.

The results of Hirata and Hoshikawa (represented by closed circles in Fig. 2) clearly show a

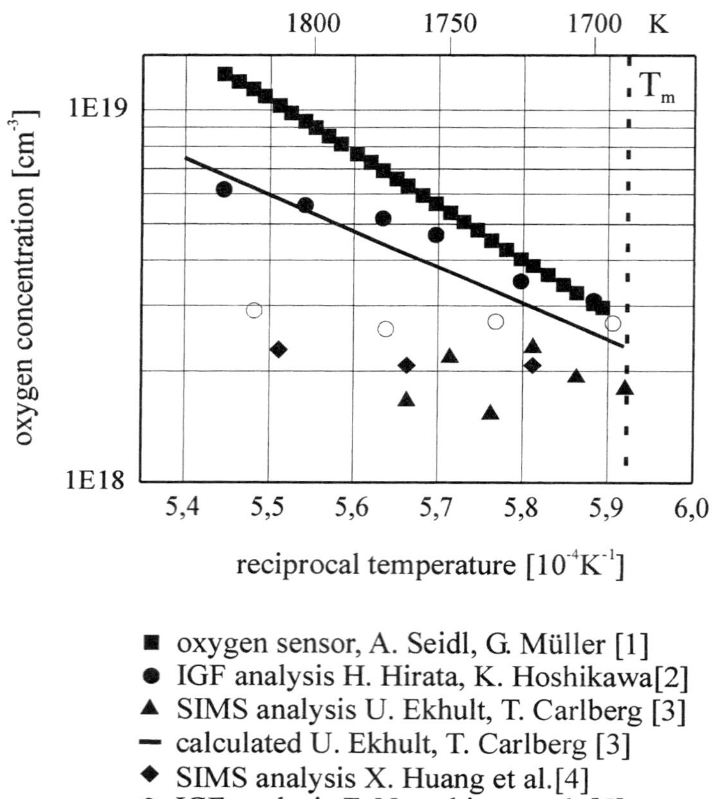

Fig. 2. Comparison of literature data for the temperature dependence of oxygen solubility in silicon melt.

temperature dependency which corresponds well to the theoretical results of Ekhult and Carlberg for the equilibrium of SiO_2 with silicon melt.

Huang et al. [4] equilibrated their specimen in a way very similar to that of Hirata and Hoshikawa. Then they analysed the specimen by IGF and SIMS, which was also calibrated using FTIR. Their results are denoted by diamond symbols in Fig. 2. It shows a very low dependency of the solubility on the temperature similar to the results of Ekhult and Carlberg. Furthermore, the authors reported that the SIMS analysis tends to give lower values for the oxygen concentration than the IGF analysis applied to the same specimen. This coincides with the fact that also Ekhult and Carlberg used SIMS analysis.

The equilibration of silicon with SiO_2 was carried out in a different way by Narushima et al. [5]. These authors kept the silicon melt in an open silica crucible and sent an oxygen flow over the melt surface, thereby establishing a thin layer of SiO_2. From this procedure they concluded to have the melt completely enclosed by SiO_2. After quenching, they analysed the specimen by IGF, which was calibrated by a set of Fe standard samples. The oxygen solubilities they found are denoted by open circles in Fig. 2. They correspond to those which Huang et al. reported to have found by the IFG technique.

One can conclude that the determination of the oxygen solubility by the described methods suffers from two problems. One is the general difficulty of absolute measurement of the oxygen concentration in a Silicon melt. This holds especially in the case if small polycrystalline samples are taken from quenched melts.

The result is a rather broad scattering of solubility data. The second problem is the uncertainty about the equilibrium, which is determining the reaction of relevance for the solubility. Several authors reported the observation of amorphous SiO at the interface between the silica crucible and the melt.

2.1.2. In-situ measurement with an oxygen sensor

As described in the introduction to this chapter, an in-situ measurement of the melt oxygen concentration is possible by the use of an electrochemical sensor. Furthermore, comparison of oxygen distributions in the bulk of the melt measured with a sensor can be used for verification of numerical models for the oxygen transport.

In Fig. 3 a schematic setup of such a sensor is shown as developed in our laboratory. The sensor is based on a solid elecrtolyte made of calcia-stabilized zirconia which acts as an oxygen ionic conductor [1]. The zirconia is formed as a tube and filled with a reference mixture of V and V_2O_3 creating the reference oxygen partial pressure $p_{O_2(ref)}$ with

$$\log p_{O_2(ref)} = \frac{-41508}{T} + 8.07 \tag{1}$$

on the inner surface of the zirconia tube [6]. On the outer surface which is immersed into the melt exists an oxygen partial pressure $p_{O_2(melt)}$ correlated to the activity a_0 by

$$a_0 = \sqrt{p_{O_2(melt)}} \exp\left(\frac{-\Delta G^0}{RT}\right) \tag{2}$$

due to the reaction

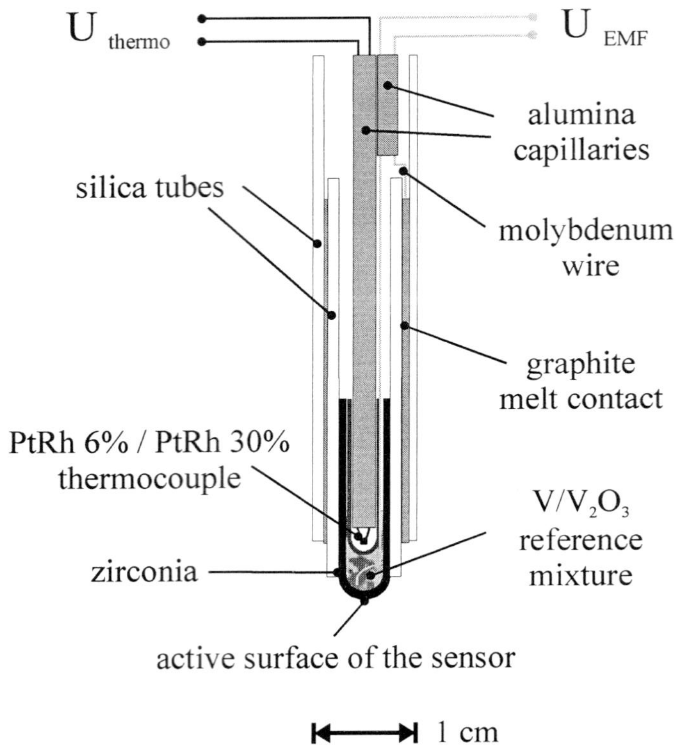

Fig. 3. Sketch of the sensor setup. The melt oxygen concentration can be calculated from the electromotive force measured between the inner and outer surface of the zirconia ionic conductor and the thermovoltage.

$$1/2\, O_2 \Leftrightarrow O_{solved} \tag{3}$$

ΔG_0 is the free energy of solution. The activity a_0 of oxygen solved in silicon melt is related to the concentration c_0 by $a_0 = f_0^* c_0$. These values are a priori unknown. For iron and other pure metal melts it is known that the activity coefficient f_0 can be set to unity for an oxygen content of less than about 0.01% [7]. As the oxygen content of silicon melts is in the range of 0.001 to 0.01%, we assume that an activity coefficient $f_0 = 1$ is also valid in this case, so that the activity a_0 equals the melt fraction of oxygen c_0. Since there are no reliable data available for ΔG_0 in the literature, this value has to be determined by calibration experiments [1].

When there is a difference between the oxygen partial pressures $p_{O_2(ref)}$ and $p_{O_2(melt)}$ on the two surfaces of the zirconia tube, diffusion of O^{2-} ions is driven through the ionic conductor until an electromotive force U_{emf} builds up which is approximately given by:

$$U_{emf} = \frac{RT}{F} \ln \left(\frac{p_e^{1/4} + p_{O_2(ref)}^{1/4}}{p_e^{1/4} + p_{O_2(melt)}^{1/4}} \right) \tag{4}$$

where R, T and F are the gas constant, the temperature and the Faraday constant; p_e is the so-called partial electronic conductivity for the ionic conductor as given by:

$$\log p_e = \frac{-68400}{T} + 21.59 \quad (5)$$

It is obvious from Eq. (4) that p_e should be small compared to the oxygen partial pressures $p_{O_2(ref)}$ and $p_{O_2(melt)}$.

Furthermore, it is favourable if $p_{O_2(melt)}$ is of the same order as $p_{O_2(ref)}$ for minimizing any kind of boundary layer reactions due to oxygen transport through the sensor and for minimizing the effect of the electronic conductivity. This can be achieved by choosing an appropriate reference mixture. A comparison of Cr/Cr_2O_3, Mn/MnO, and V/V_2O_3 revealed the latter one to be optimum for the application in Czochralski silicon melts [8,9].

Since all relevant reactions depend strongly on temperature, the sensor contains a B-type thermocouple protected by an alumina tube. Once the electromotive force and the temperature are known, the melt oxygen partial pressure $p_{O_2(melt)}$ can be calculated using Eqs. (1), (4), and (5). During this procedure, an error is induced due to the variations of temperature [10]. This error can be minimized by measuring the temperature as close as possible to the ionic conductor and the reference mixture. An appropriate design to achieve this objective is shown in Fig. 3.

It seemed to be desirable to test the sensor under conditions comparable to the literature data available for the solubility of oxygen in silicon melt. For that purpose 100 g of silicon melt were kept in a closed silica crucible. When the melt was believed to be equilibrated with the ambient silica and SiO gas, the signals of an oxygen sensor immersed into the melt were recorded under slow variation of the melt temperature. In Fig. 2 the melt oxygen concentrations evaluated from the sensor voltages are plotted against $1/T$ as squares. It can be seen that a reasonable value for the oxygen solubility close to the melting point is found. A comparison of the temperature dependence of the solubility resulting from the value of ΔG_0 [8] with the other plotted data is rather difficult, since the FTIR method used for the calibration of the sensor differs from the methods used by the other authors.

However, from a practical point of view, it is satisfactory for the application if an electrochemical oxygen sensor is capable to predict the oxygen concentrations that will be found in Czochralski crystals by an in-situ measurement of the oxygen concentration in the melt, which will be treated in the next chapter.

2.2. O-concentration in Si under Czochralski conditions

In order to test an oxygen sensor under Czochralski conditions, it was dipped into the melt of a Czochralski puller during a 4″ crystal growth. Fig. 4 shows a comparison of the oxygen concentrations in the melt measured by the sensor and the results of FTIR analysis at several positions of the grown crystal, denoted by squares. The radial positions of the analysed crystal samples are close to the periphery of the crystal. The axial positions correspond to that time of the growth run when the portion of the crystal was grown. The oxygen concentration found in the specimen from the crystal differs from that measured in the melt by a constant factor of about 1.3–1.4.

This result was very encouraging with respect to a possible control of the oxygen concentration in the crystal assisted by in situ sensor measurements. However, it remained to be clarified what was the reason for the deviation of the oxygen concentrations measured in the crystal and in the melt. For that reason, various extended scans of the oxygen distribution during one crystal growth run were performed. The sensor is mounted on the lower end of a steel tube, which can be inserted into any

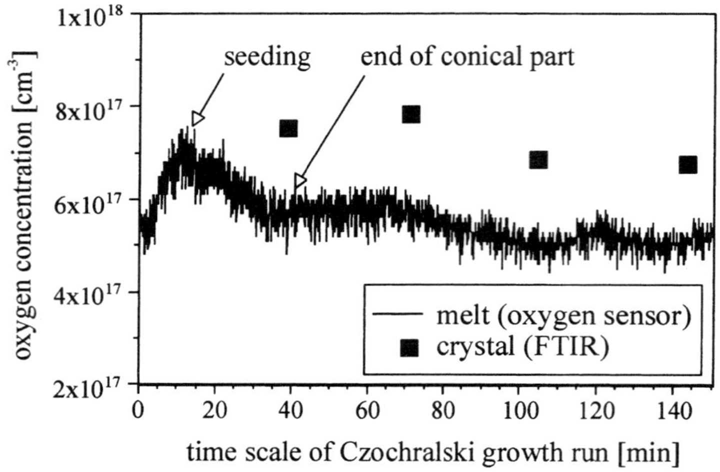

Fig. 4. Comparison of oxygen concentrations in the melt measured by an oxygen sensor during a Czochralski growth run to the oxygen concentrations found in the corresponding crystal sections by FTIR.

industrial Czochralski puller. The vertical and radial positioning of the sensor in the melt is provided by means of a motion feed-through. As the melt height is continuously decreasing during growth, a quick performance of the scans is important to achieve 'momentary images'.

In Fig. 5 a typical oxygen distribution resulting from this procedure is shown. One can see that the oxygen concentration close to the bottom of the crucible differs from that close to the melt surface and correlates to the oxygen concentration found in the crystal. This was again confirmed by FTIR analysis. It might indicate that oxygen-rich melt from the crucible bottom is transported to the growth

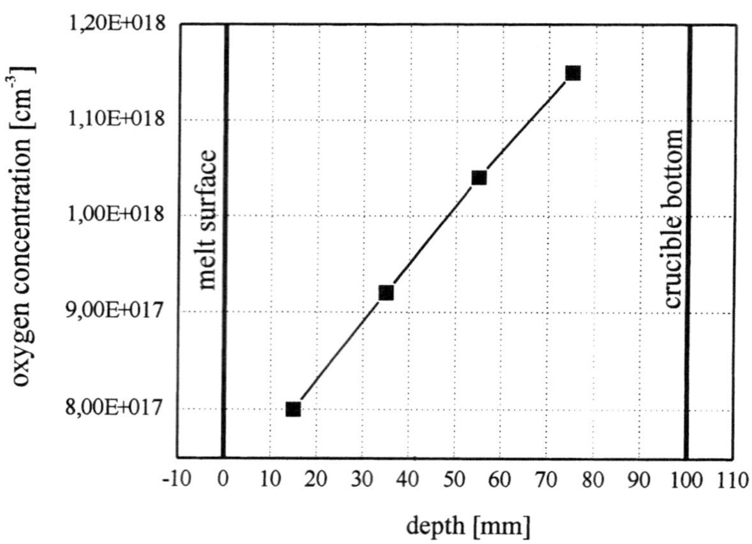

Fig. 5. Vertical profile of oxygen concentration in the bulk of the melt in a Czochralski crucible at a radial position $r=0.5*R_c$ during a Czochralski growth run.

interface by the convective flow. This confirms the importance of both experimental and numerical examination of the melt flow structure during the Czochralski growth process for the understanding of the oxygen transport.

3. Numerical modelling of oxygen transport in silicon Czochralski melts

3.1. Model conception

In order to control the oxygen transport during a Czochralski growth process a simulation model is needed which can predict the oxygen distribution at the crystal–melt interface in dependency from the process parameters. A numerical process model has to consider all relevant physical and physico-chemical mechanisms which were discussed in Sections 1 and 2.

(i) oxygen source (dissolution of crucible wall)

(ii) oxygen transport in the silicon melt by coupled convective–diffusive transport including buoyancy, rotation, capillary and eventually magnetic fields.

(iii) oxygen drain (evaporation of SiO from the free melt surface and transport by Ar-flow).

We know from Section 2 that the source mechanism (i) is strongly temperature dependent.

The convection phenomena in (ii) and (iii) are at least containing two mechanisms which are driven by the temperature field (buoyancy, thermo-capillary). The Ar-flow (iii) is also partly driven by thermal buoyancy. On the other hand, the solutal effect coming from the oxygen concentration in the melt or Ar-gas can be neglected compared to the thermally induced density variation. Thus, it can be concluded that for a calculation of the flow velocity field — both in the melt and in the gas phase — only the transport of heat has to be considered, whereas the transport of oxygen can be neglected. With respect to the differential equations this means, that the heat transport equations can be solved separately from the decoupled species transport equation. A model to describe the global heat transfer in a typical Czochralski apparatus that accounts for radiative, diffusive and convective heat transport is published elsewhere (e.g. STHAMAS [11]).

The calculation of the defect transport, defect formation, and precipitation in the growing Si crystal can also be treated independently and decoupled from the transport mechanisms, however, by using the calculated thermal field.

3.2. Model for oxygen transport and boundary conditions

While oxygen occurs only in low concentration in molten silicon (<100 ppm) and in the gas ($<0.3\%$) it can can be described with the standard equation (2nd Fick's law) equivalently by using the solution of the velocity field $\boldsymbol{v}$ from the heat transfer problem:

$$\frac{\partial C}{\partial t} = -v_i \frac{\partial C}{\partial x_i} + \frac{\partial}{\partial x_i} D \frac{\partial}{\partial x_i} C + q_C \tag{9}$$

where C is the concentration and D the diffusivity of oxygen. q_C describes sources and drains of oxygen.

3.2.1. Boundary conditions

According to Fig. 1 we have the following boundary conditions for the oxygen transport. Oxygen enters into the silicon melt according to the dissolution rate of the crucible which depends on the temperature and the adjacent melt flow along with other factors.

This implicits as boundary condition for oxygen at the silica crucible wall with an oxygen flux as a function of temperature and velocity field:

$$(\nabla C)\boldsymbol{n} = F(T, \boldsymbol{v})$$

where $\boldsymbol{n}$ is the unit vector perpendicular to the melt–crucible interface. This dependence is not well known, but it can be assumed that the oxygen concentration at the melt–crucible interface is well described by the temperature dependent solubility due to the fast kinetics of the dissolution process. Kim et al. [12] showed that in their simulation this assumption leads to dissolution rates in agreement with measurements within the limits of error. The evaporation of the silicon monoxide at the melt/gas surface is also a process with fast kinetics. The rate determining step is the relatively slow transfer of oxygen via advection and diffusion from inside the melt to the surface. Therefore, silicon monoxide evaporates almost instantaneously upon reaching the free surface and may be set to zero at the free surface. The advantage of this approximation is huge since now the oxygen transport in the Ar-gas phase needs not to be calculated. Heat transfer by convection of the Ar-gas can eventually be estimated by a simplified engineering model (e.g. heat transfer coefficient, effective heat conductivity).

Nevertheless, a small, but negligible amount of oxygen is existing at the free surface. Huang et al. [13] measured temperature dependent evaporation rates of silicon monoxide, which were used by Togawa et al. [14] to calculate the oxygen distribution in the silicon melt. This calculation leads to an oxygen concentration at the free melt surface which is approx. 1% of the oxygen concentration at the crystal–melt interface.

Since in this model the concentration at the free surface near the triple point is not fixed to be zero, a better reproduction of the experimentally observed radial non-uniformity of the oxygen distribution in the vicinity of the crystal rim could be achieved. At the growing interface the segregation coefficient k close to unity leads to a 'no-flux' boundary condition, as no oxygen is rejected ($k<1$) or deflected ($k>1$):

$$(\nabla C)\boldsymbol{n} = 0$$

where $\boldsymbol{n}$ is the unit vector perpendicular to the melt–crystal interface.

3.3. State of the art of the simulation of the oxygen transport

The flow in a real Czochralski apparatus as used in industrial processes of silicon growth is 3-dimensional, turbulent and time dependent. Due to the relatively small thermal mass of the silicon melt in the whole furnace, temperature field and stress in the silicon crystal may be predicted sufficiently precise by an approximate treatment of the heat transport in the melt [15,16].

However, in order to predict the oxygen incorporation into the growing crystal, i.e. the transport of oxygen to the crystal–melt phase boundary, it is necessary to consider the convective flow in the silicon melt. The Schmidt number $Sc = \nu / D$ of oxygen in silicon melt is three orders of magnitude larger than the Prandtl number Pr $= \nu/\chi$ (ν = kinematic viscosity, χ = thermal diffusivity). Thus, the transport of oxygen is significantly more sensitive on convection than the heat transfer. In order to cope with the turbulent flow structures in the Si–melt various turbulence models have been applied [15,17]. Most of the models are using radial symmetry (called 2D); 3D treatments are rather rare.

The high Schmidt number of oxygen transport in silicon melt in the range from 6 to 9 leads to a quite narrow concentration boundary layer at the crucible–melt interface in comparison to the thermal and velocity boundary layers of the flow field [17].

This small concentration boundary layer determines the size of the numerical grid elements which are required to resolve the concentration profile in the layer, which is necessary to get a grid independent solution. Because it is not possible and not necessary to have such a fine grid in the whole melt volume, the numerical procedure has to cope with highly non-uniform grids. The computer code STHAMAS [11], which was developed in our group for global simulation of the Czochralski process, was used to analyse the problem of a resolution of the oxygen boundary layer along the crucible wall. For that purpose we first carried out a global simulation (steady, axisymmetric) of the heat transport and the temperature field in a typical Cz silicon puller. The crucible diameter is about 30 cm (20 kg silicon in the melt); the crystal diameter is about 10 cm. The convection in the melt was treated by a standard k–ϵ turbulence model. Heat transfer in the Ar-gas was considered only by diffusion and radiation according to the arguments given above, although our code STHAMAS is also capable to treat turbulent convection in the gas. The heating power was adjusted to fit the melting temperature T_{melt} at the triple point. For the calculation of the oxygen transport in the melt we used the averaged velocity field from the global thermal simulation. Now we solved the turbulent diffusion–convection equation for the oxygen transport with the following boundary conditions. The oxygen concentration at the melt–crucible interface was set to the temperature dependent solubility value, to zero at the melt–gas interface, and a segregation coefficient of unity was taken at the melt–crystal interface. The results of the calculated oxygen concentration profiles in the vicinity of the crucible wall are shown in Fig. 6 for three different sizes of the numerical grid. The thickness of the concentration boundary layer was found to be about 0.5 mm. It follows clearly that a minimum of 5 grid points in the first 0.5 mm is needed to resolve the concentration profile within the boundary layer. In the bulk of the melt the used grid size can be 40 times larger.

It should also be pointed out that the small dimensions of the fine grid are only needed perpendicular to the crucible surface. Parallel to the crucible surface the grid may also be taken as coarse as in the bulk.

In the above test calculations we have undertaken no effort to verify the accuracy of the turbulent flow model by comparison with measured temperature or oxygen distribution inside the melt. However, we have tested our model to make a calculation of the oxygen distribution in the melt for a CZ-geometry which was studied by the oxygen sensor (compare to Section 2). The calculations were performed in the same way as described before, except that for the crucible–melt interface a temperature dependent equilibrium oxygen concentration was used [13] as boundary condition:

$$C(T) = 1.32*10^{19} \exp(-3200/T).$$

The calculated flow field and the oxygen distribution are shown in Fig. 7. Due to the high Schmidt number, the concentration distribution is strongly influenced by the melt convection. It can be seen

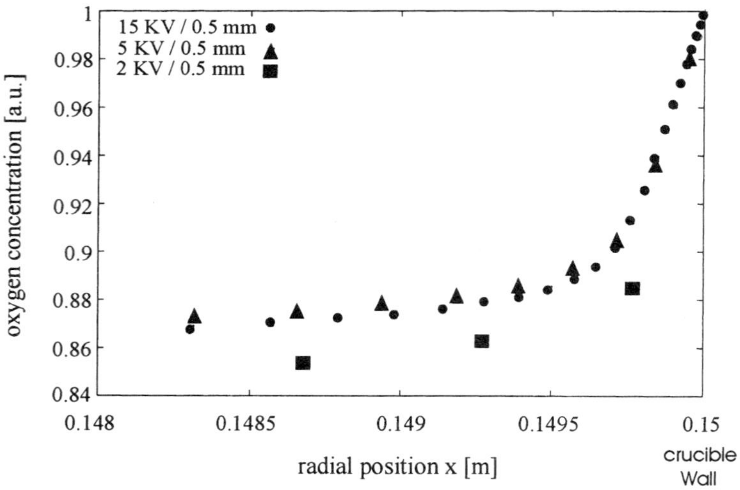

Fig. 6. Numerically calculated oxygen concentration in the vicinity of the vertical crucible wall ($x = 0.15$ m) where the oxygen concentration is set to unity versus radial position x at a vertical position of 50% of the melt height. The different symbols belong to different numerical grids. KV means the number of control volumes. The width of the boundary layer is of the order of 0.5 mm. It is shown that the coarse grid is not sufficient to resolve the boundary layer. A minimum of 5 grid points is needed within the 0.5 mm to resolve the boundary layer.

that the calculated overall oxygen concentration is about 20 percent higher than found in the experiment. Furthermore, the calculated concentration gradient of $2*10^{16}$ cm^{-4} in the bulk of the melt is significantly lower than the measured one ($1*10^{17}$ cm^{-4}). Calculations by Togawa et al. [18], which were performed using a model for the SiO evaporation rate, also lead to higher oxygen concentrations in the melt and crystal, while the concentration gradients are matched closer with our experimental results.

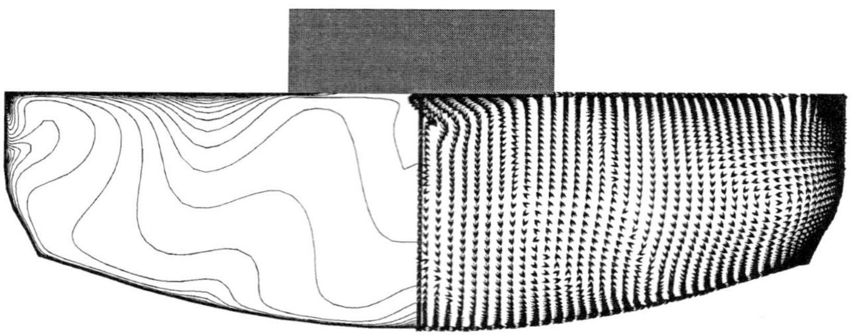

Fig. 7. Results of numerical simulation of the oxygen transport in a silicon Czochralski melt ($R_c = 14'$) with a 4′-crystal using a turbulence model and boundary conditions as given in the text, flow field (right) and lines of equal oxygen concentration (left) with a distance of 0.01 C_{max}. The rotation rates are $\Omega_c = 5$ s^{-1} for the crucible and $\Omega_x = -20$ s^{-1} for the crystal.

4. Conclusions and perspectives for oxygen control

The available electrochemical oxygen sensor can be used to study in situ the relation of the oxygen concentration in the melt in dependence from any parameter variations including magnetic fields.

The numerical process simulation of the oxygen transport under consideration of parameter variations including magnetic fields is at present in an early stage and less developed than the global heat transport modelling. One reason for that is the necessity of using adequate modelling of turbulent flow. The second one is that the concentration boundary conditions are much more sensitive to the transport conditions than in the case of the heat transport. Thus the experimental analysis of oxygen transport will continue to play its important role in the next years. It cannot be replaced at present by numerical simulations. Nevertheless it is highly desirable to improve the status of numerical modelling by finding more adequate turbulence models for the CZ process.

Acknowledgements

The authors wish to thank A. Seidl and J. Fainberg for their work on several topics involved in this paper. The financial support by the Bavarian research consortium FORTHWIHR and the Bavarian Ministry of Economy, Transport and Technology is gratefully acknowledged.

References

[1] A. Seidl, R. Marten, G. Müller, J. Electrochem. Soc. 141 (1994) 2564.
[2] U. Ekhult, T. Carlberg, J. Electrochem. Soc. 136 (1989) 551.
[3] H. Hirata, K. Hoshikawa, J. Cryst. Growth 106 (1990) 657.
[4] X. Huang, K. Terashima, H. Sasaki, E. Tokizaki, S. Kimura, Jpn. J. Appl. Phys. 32 (1993) 3671.
[5] T. Narushima, K. Matsuzawa, Y. Mukai, Y. Iguchi, Materials Transactions, JIM 35 (1994) 522.
[6] I. Barin, Thermochemical Data of Pure Substances, VCH, Weinheim, 1989.
[7] A. Weyl, S.W. Tu, D. Janke, Steel Research 65 (1994) 167.
[8] A. Seidl, R. Marten, G. Müller, J. Electrochem. Soc. 144 (1997) 3243.
[9] A. Seidl, R. Marten, G. Müller, J. Crystal Growth 166 (1996) 680.
[10] K. Kakimoto, H. Naguchi, M. Eguchi, J. Electrochem. Soc. 144 (1997) 4045.
[11] J. Fainberg, H.-J. Leister, G. Müller, J. Crystal Growth 180 (1997) 517.
[12] K.M. Kim, W.E. Langlois, J. Electrochem. Soc. 138 (1991) 1850.
[13] X.M. Huang, K. Terashima, H. Sasaki, E. Tokizaki, Y. Anzai, S. Kimura, Jap. J. Appl. Phys. 33 (1994) 1717.
[14] S. Togawa, X. Huang, K. Izunome, K. Terashima, S. Kimura, J. Crystal Growth 148 (1995) 70.
[15] T.A. Kinney, R.A. Brown, J. Crystal Growth 132 (1993) 551.
[16] E. Dornberger, E. Tomzig, A. Seidl, S. Schmitt, H.-J. Leister, Ch. Schmitt, G. Müller, J. Crystal Growth 180 (1997) 461.
[17] F. Rosenberger, G. Müller, J. Crystal Growth 65 (1983) 91.
[18] S. Togawa, K. Izunome, S. Kawanishi, S.-I. Chung, K. Terashima, S. Kimura, J. Crystal Growth 165 (1996) 362.

ELSEVIER

Microelectronic Engineering 45 (1999) 149–154

MICROELECTRONIC ENGINEERING

Vacancy distribution measurements in CZ Si crystals grown by different pulling rate

Yukio Takano*, Masaru Sutoh, Ayumu Satoh

Department of Materials and Technology, Science University of Tokyo, 2641 Yamazaki, Noda-shi, Chiba, 278, Japan

Abstract

Vacancy concentration distribution is estimated by measuring deep levels induced by N_2-vacancy complexes in the wafers which are prepared from Si crystals grown by different pulling rate. It is found that the vacancy concentration is very low inside the OSF-ring and a little low in the ring region. Two kinds of the vacancy density profiles are observed in the specimens without the OSF-ring which are prepared from Si crystals grown by fast (1.4 mm/mm) and slow (0.4 mm/mm) pulling rate: one is a gradual and the other is a abrupt zigzag profile. We interpret the gradual profile is caused by macroscopic density change of the octahedral void defects and the abrupt variation is accompanied with the growth striation.

Keywords: Estimation of vacancy concentration; Vacancy density profiles; Effect of crystal pulling rate on vacancy density

1. Introduction

Octahedral void type defects [1] called as COP (Crystal Originated Particles) [2], FPD (Flow Pattern Defects) [3] and LSTD (Light Scattering Tomography Defects) [4] are known to be generated in as grown CZ crystals and degrade the gate oxide integrity of MOS devices [5]. Generation of the void defects are considered to be closely related with supersaturation of vacancies in the crystals. However, there is few experimental data about the vacancy, because it is very difficult to measure the vacancy density precisely.

A few years ago, we found that deep levels were introduced when Si wafers were annealed in N_2 and then cooled rapidly to room temperature [6]. The deep levels (Ec−0.5 eV) were confirmed to be induced by N_2 and vacancy complexes [7]. In this report, we tried to estimate the vacancy density in Si wafers by measuring the deep levels.

2. Experimental

Specimen wafers 1–3 were prepared from three kinds of n-type, 8–10 Ω-cm, (100), 100 mm in

*Corresponding author. Tel.: +81-471-241-501; fax: +81-471-239-362.
E-mail address: takano@rs.noda.sut.ac.jp (Y. Takano)

0167-9317/99/$ – see front matter
PII: S0167-9317(99)00117-3

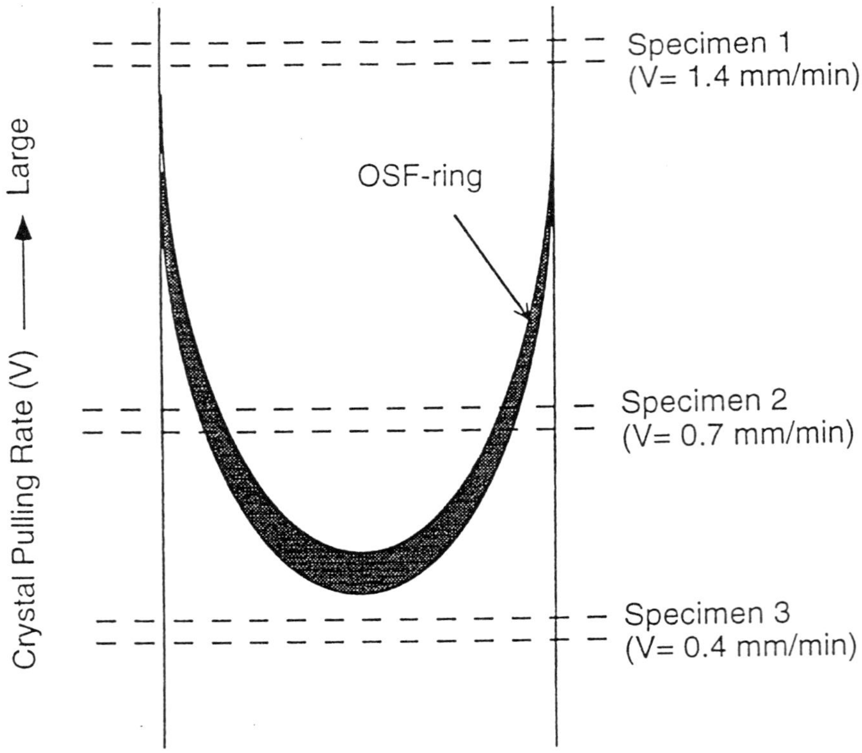

Fig. 1. Schematic drawing of the relationship between the OSF-ring and the crystal pulling rate of the specimens.

diameter CZ crystals. The crystal pulling rate of them were 1.4, 0.7 and 0.4 mm/min, respectively. Relationship between the pulling rate and the oxidation-induced staking fault (OSF)-ring is schematically shown in Fig. 1. It is known that the OSF-ring diameter becomes smaller as the crystal pulling rate decreases and finally vanishes. Inside and outside of the OSF-ring is believed to be a vacancy rich and an interstitial rich region, respectively [8]. This is because a lot of the octahedral void defects have been observed inside of the OSF-ring, on the other hand, interstitial dislocation

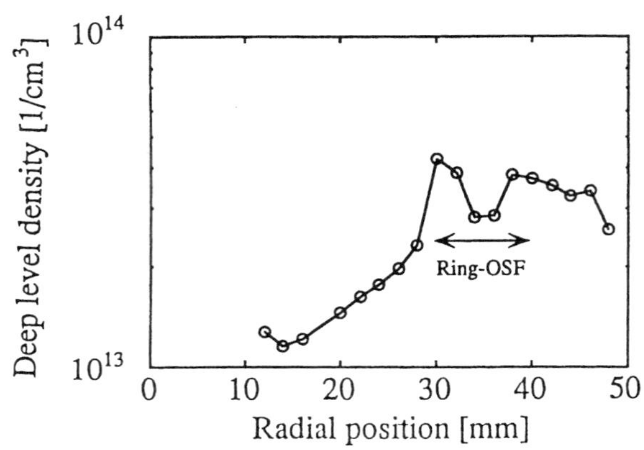

Fig. 2. Distribution of the deep level density along a diameter of the specimen 2 (v=0. 7 mm/mm).

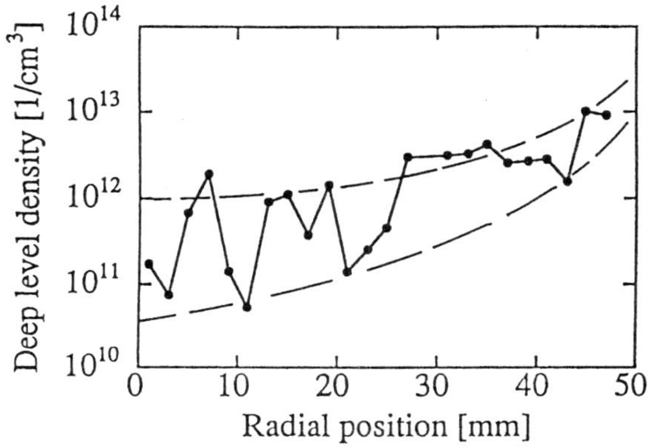

Fig. 3. Distribution of the deep level density along a diameter of the specimen 1 which was prepared from a crystal of $v = 1.4$ mm/min.

loops have been observed outside of it. In the OSF-ring region, many OSFs are observed when the specimen is oxidized at high temperatures such as 1200°C.

All of the specimen wafers were annealed at 1000°C for 3 mm in N_2 and O_2 mixed gas and then quenched to room temperature. N_2 diffuses into the crystals through the growing SiO_2, however, the vacancies cannot diffuse in in the crystal, since excess interstitials supplied from the growing SiO_2 suppress the vacancy diffusion. The deep levels induced by N2-vacancy complexes are proportional to the vacancy concentration in the specimen annealed under the above condition, because the vacancy concentration is lower than that of diffused N_2 [9].

3. Results and discussions

Fig. 2 shows the distribution of the deep level density along the wafer diameter of the specimen 2. In Fig. 2, the remarkable change of the deep level density is observed: the vacancy concentration is very low in the central region and a little low in the OSF-ring region. It has already shown that many

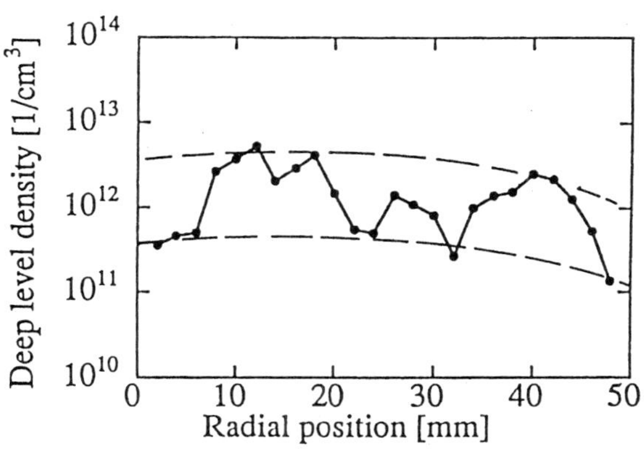

Fig. 4. Distribution of the deep level density along a diameter of the specimen 3 ($v = 0.4$ mm/mm).

void defects are observed inside the OSF-ring. Therefore, the vacancy concentration should be lower inside the OSF-ring, since the vacancies should consume for the void defects to grow up further. In the OSF-ring region, there are large SiO_2 precipitate nucleus which can supply excess interstitials from SiO_2/Si interface and annihilate the vacancies. This is why the vacancy density is lower inside and in the OSF-ring region.

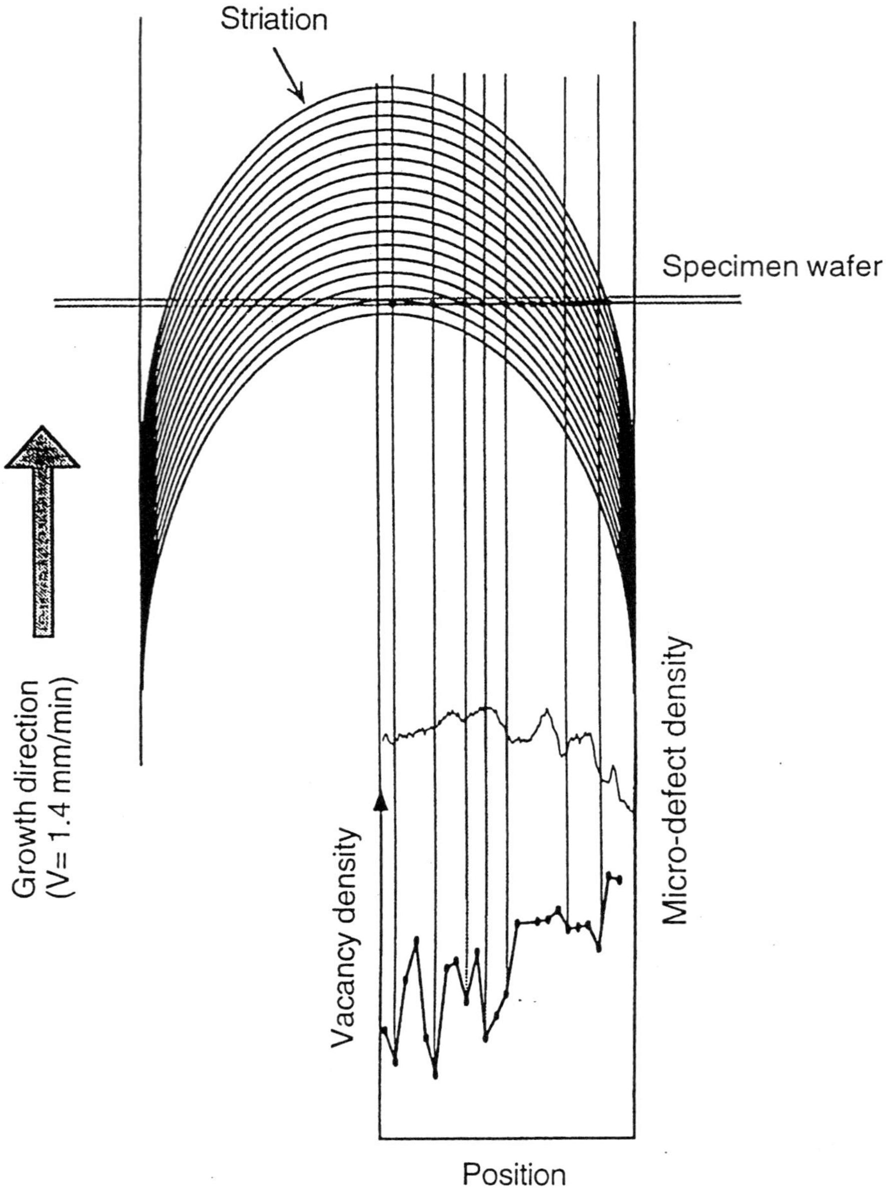

Fig. 5. Schematic drawing of growth striation along directions parallel and perpendicular to the crystal growth direction, vacancy and micro-defect density distribution along a diameter of the specimen 1.

Figs. 3 and 4 show the distribution of the deep level density in the specimen 1 and 3, respectively. Two kinds of the density profiles are superposed in both of the figures: one is the gradual and the other is the abrupt zigzag profiles. The vacancy density is gradually increasing towards the wafer periphery in Fig. 3. The distribution is also understandable if we accept the vacancy consuming for the void defects growth as shown in the case of Fig. 2. In the wafer, the void defects are formed in the whole region except near the periphery in which the vacancies easily diffuse out from the crystal surface.

The vacancy distribution of Fig. 4 seems to be similar to that of the supersaturation of the vacancy after the crystal growth. In the specimen 3, silicon interstitial density is larger than that of the vacancy since the specimen is prepared from the outside of the OSF-ring. Therefore, the vacancy distribution is maintained to be that of the supersaturation.

Next we will explain the zigzag profile of the vacancy density shown in Figs. 3 and 4 to be due to growth striation such as oxygen of the crystal. Fig. 5 schematically shows the growth striation along parallel and perpendicular to the growth direction in the specimen 1 (the pulling rate $v = 1.4$ mm/min). The results of Fig. 3 is also shown to compare with the striation in Fig. 5. It is noted that the period of the the zigzag patter roughly coincides with that of the growth striation.

Relationship between the micro-defect and the vacancy density was also studied in the specimen prepared from the same crystal ($v = 1.4$ mm/min) as the specimen 1. The specimen was annealed at 800°C for 4 h and 1000°C for 16 h. The micro-defect density was estimated from an X-ray topograph taken from the specimen as shown in Fig. 6. The micro-defect and the vacancy density profiles are found to be closely related with each other as shown in Fig. 5. The void defects are known to be formed during the crystal growing at high temperatures between 1050°C and 1100°C. In that temperature range, we consider the vacancies which make complexes with interstitial oxygen atoms

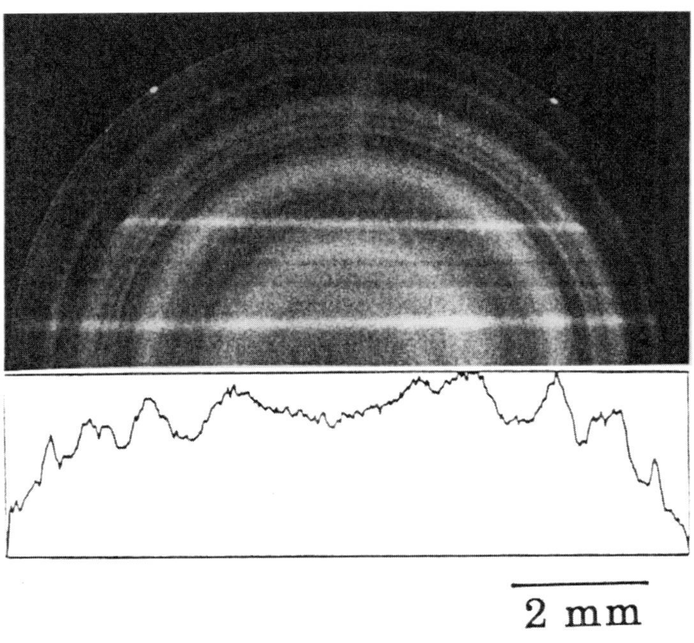

Fig. 6. X-ray topograph of the specimen prepared from the same crystal as the specimen 1 after annealing at 1200°C for 1 h.

don't move to form the void defects and make the precipitate nucleus in other temperature range. In order to confirm this assumption, further investigation is necessary of course.

4. Conclusion

We estimate vacancy concentration distribution by measuring the N_2 related deep levels in wafers which are prepared form Si crystals grown by different pulling rate. In the wafer containing OSF-ring, it is found that the vacancy concentration is very low in a central region and a little low in the OSF-ring region. Two kinds of the vacancy density profiles are observed in the wafers without the OSF-ring prepared from Si crystals grown by fast (1.2 mm/min) and slow (0.4 mm/min) pulling rate: one is the gradual and the other is the abrupt variation.

We explain the profiles by considering density change of the void defects and the growth striation.

References

[1] M. Kato, T. Yoshida, Y. Ikeda, Y. Kitagawara, Jan. J. Appl. Phys. 35 (1996) 5597.
[2] J. Ryuta, E. Morita, T. Tanaka, Y. Shimanuki, Jan. J. Appl. Phys. 29 (1990) L1947.
[3] H. Yamagishi, I. Fusegawa, N. Fujimaki, M. Katayama, Semicond. Sci. Technol. 7 (1992) A135.
[4] P. Gall, J.P. Fillard, J. Bonnafe, T. Rakotomavo, H. Rufer, H. Schwenk, Defect Control in Semiconductors, in: North-Holland, Amsterdam, 1990, p. 255.
[5] M. Iitsumi, M. Tomita, M. Yamawaki, J. Appl. Phys. 78 (1995) 1940.
[6] N. Fuma, K. Tashiro, K. Kakumoto, Y. Takano, Mater. Sci. Forum (1995) 196–201, Also Trans. Tech. Pub. Switzerland (1995) 797.
[7] N. Fuma, K. Tashiro, K. Kakumoto, Y. Takano, Jpn. J. App. Phys. 35 (1996) 1993.
[8] S. Sadamitsu, S. Umeno, Y. Koike, M. Hourai, S. Sumita, T. Shigematsu. Jpn. J. Appl. Phys. 32 (1993) 3675.
[9] K. Kakumoto, Y. Takano, Proceedings of The 2nd International Symposium on Advanced Science and Technology of Silicon Materials, in: The 145th Committee on Processing and Characterization of Crystals, 1996, p. 437.

ELSEVIER

Microelectronic Engineering 45 (1999) 155–160

MICROELECTRONIC ENGINEERING

Uniform precipitation of oxygen in large diameter wafers

G. Kissinger[a,*], J. Vanhellemont[b], U. Lambert[b], D. Gräf[b], H. Richter[a]

[a]Institute for Semiconductor Physics, Walter-Korsing-Str. 2, 15230 Frankfurt (Oder), Germany
[b]Wacker Siltronic AG, P.O. Box 1140, 84479 Burghausen, Germany

Abstract

In the present stage of development, 300 mm crystals often contain a transition from vacancy-rich to interstitial-rich. Due to this radially varying concentration of intrinsic point defects, the radial size distribution of grown-in oxide precipitate nuclei is also inhomogeneous in these wafers. In order to achieve a radially uniform bulk defect density for internal gettering, a slow temperature ramp induced growth of all grown-in oxide precipitate nuclei is the appropriate procedure to overcome problems resulting from the inhomogeneous size distribution of grown-in nuclei. This approach is based on the observation that in contrast to the nuclei size distribution, the nuclei density distribution is homogeneous over the wafer, independent of the dominant intrinsic point defect.

Keywords: Oxide precipitation; 300 mm Si wafers; Defect density

1. Introduction

The silicon community has accepted the challenge of manufacturing and processing 300 mm wafers. This imposes great demands on the development of equipment to ensure uniformity of temperature, doping, and structuring over more than double the area of the present industry standard, 200 mm wafers. Uniformity of oxygen precipitation, a high density of gettering bulk defects, and an excellent denuded zone are some of the other major requirements to be met through the interaction between defect engineering and crystal growth.

Most problems connected with crystal defects in 300 mm silicon wafers are a continuation of those for wafers of smaller diameter. In recent years interest has focused on D-defects, being octahedral vacancy agglomerates about 30–200 nm in diameter [1,2]. They are important because they cause the breakdown of thin gate oxides and thus degrade the device yield [3]. However, although they have been present in silicon wafers of lower diameter for a long time [4], their exact nature was only discovered during the last few years [1].

Another important type of defects are the grown-in oxide precipitate nuclei in Czochralski (CZ) silicon. Their grown-in size distribution is strongly influenced by intrinsic point defects, their

*Corresponding author. Tel.: +49-335-56250; fax: +49-335-5625-300.

0167-9317/99/$ – see front matter
PII: S0167-9317(99)00119-7

supersaturation, and precipitation behavior during crystal cooling [5]. Oxide precipitate nuclei are also important for the device yield because they can grow during device processing and form bulk defects, which are necessary for gettering heavy metal impurities. However, the oxide precipitate nuclei can also cause harmful denuded zone defects in the active device area [6]. These phenomena are determined by the interaction of crystal growth conditions with the thermal conditions of subsequent device processing [7].

2. Trends for crystal defects in connection with the transition to 300 mm crystals

In the present stage of development, 300 mm crystals often contain a transition from vacancy-rich to interstitial-rich which is not located at the perimeter of the crystal. Due to this radially varying concentration of self-interstitials and vacancies, the radial size distribution of grown-in oxide precipitate nuclei is also inhomogeneous in these wafers. The bulk defect density can therefore change markedly with the radial position, especially in the region of the transition from vacancy-rich to interstitial-rich.

The cooling rate of the crystals further decreases with the transition to 300 mm crystals [8]. This is the result of the larger volume of the crystals, but also necessary to prevent dislocation formation during crystal growth. A low cooling rate results in small grown-in oxide precipitate nuclei because the lower the cooling rate of the crystals, the more vacancies are annihilated at D-defects prior to oxide precipitate nucleation during crystal cooling below 900°C. With decreasing vacancy supersaturation, the maximum nucleation rate of oxide precipitate nuclei shifts to lower temperatures, and the critical radius of the nuclei decreases [5]. The problem that results from the small number of grown-in oxide precipitate nuclei in 300 mm wafers is that it is more difficult to obtain enough bulk defects for internal gettering by oxygen precipitation under the usual processing conditions for wafers with smaller diameter.

The size and density of the D-defects also depend on the cooling rate of the crystal. Low cooling rates lead to larger D-defects of lower density, and thus it is expected that the gate oxide integrity will improve with the transition to 300 mm crystals. However, making the D-defects electrically inactive with respect to the gate oxide quality by internal void oxidation in high temperature anneals [9] will need more time for 300 mm wafers because of the larger size of the grown-in voids.

In this paper, we show that it is possible to obtain a uniform and high density of gettering bulk defects in 300 mm wafers, even if they are obtained from a crystal with a transition from vacancy-rich to interstitial-rich. This can only be achieved if the total density of grown-in oxide precipitate nuclei on each position of the wafer is grown by a slow temperature ramp starting from 500°C to form large bulk defects. This approach is based on the observation that in contrast to the nuclei size distribution, the nuclei density distribution is uniform across the wafer, independent of the dominant intrinsic point defect.

3. Experimental

Boron doped CZ (001) silicon wafers with a diameter of 300 mm, a concentration of interstitial oxygen of 6.3×10^{17} cm^{-3} (DIN 50438/1), and a resistivity of 5–10 Ωcm were used for the experiments.

All defect densities (N_{LST}) were measured by infrared light scattering tomography (IR-LST) using a Ratoc MILSA-IRHQ2 light scattering tomograph equipped with a Nd-YAG laser of 1060 nm wavelength.

In order to obtain information about the grown-in size distribution of oxide precipitates, defect density spectra were measured by a combined LST/annealing method, which is described in detail elsewhere [7,10]. The spectra show the density of grown-in oxide precipitate nuclei (N_{LST}) as a function of their stability temperature, which is related to the critical radius of the oxide precipitate nuclei. The critical radius of precipitate nuclei can be derived from the homogeneous nucleation theory (see e.g. Ref. [11]). An increasing stability temperature is tantamount to an increasing radius of the grown-in precipitate nuclei.

4. Results and discussion

The supersaturation of intrinsic point defects varies markedly from center to edge for crystals with a transition from vacancy-rich to interstitial-rich. In the vacancy-rich central part, the supersaturated vacancies agglomerate during cooling of the crystals by forming D-defects in the temperature range from 1050°C to 900°C [12]. However, in the transition region from the vacancy-rich central part to the interstitial-rich outer part of the crystals, the vacancy supersaturation remains subcritical leading to the region of the highest residual supersaturation before oxide precipitate nucleation starts below 900°C in the crystal cooling process. This results in the largest grown-in oxide precipitate nuclei because the nucleation rate is clearly dependent on the supersaturation of intrinsic point defects [13]. The smallest nuclei usually exist in the interstitial-rich outer part of the crystals [5]. Fig. 1 shows defect density spectra of characteristic regions of the 300 mm wafers studied. As predicted, the interstitial-rich region in the outer wafer part has the smallest nuclei with the lowest stability temperature (10 mm from wafer edge). The vacancy-rich central region (85 mm from wafer edge) contains nuclei of medium size, because a high concentration of vacancies was agglomerated before oxide precipitate nucleation. The largest grown-in nuclei (about 10^8 cm^{-3}, marked by the arrow) were found in the transition region exactly at the radial position 45 mm from wafer edge. Here, a stacking fault ring can be observed on the wafers after subsequent wet oxidation at 1100°C for 2 h. The largest nuclei of the wafer, present in this region, first exceed the critical strain for stacking fault generation during processing [14]. The large nuclei stable at 900°C were formed in an earlier stage of oxide precipitate nucleation and have reduced the residual concentration in the center of the peak, before again nuclei in a higher density are formed in the main nucleation stage. Often marginal bands of large grown-in nuclei are observed on both sides of the stacking fault ring region. They also result from the residual peak of high vacancy supersaturation, but the relatively large nuclei are formed in the main nucleation stage, see e.g. the defect density spectrum 60 mm from wafer edge. In summary, Fig. 1 shows that the grown-in size distribution of oxide precipitate nuclei varies markedly. However, the total density of nuclei is nearly constant across the wafer. Thus, the only way to obtain a laterally uniform bulk defect density is to grow all nuclei to large stable bulk defects. This has the additional advantage that the maximum bulk defect density for internal gettering is obtained.

Grown-in oxide precipitate nuclei can only grow during ramping of wafer processing if their growth rate is high enough that they can remain supercritical in size, because the critical radius of oxide precipitate nuclei increases exponentially with the temperature. This can only be ensured by a ramp

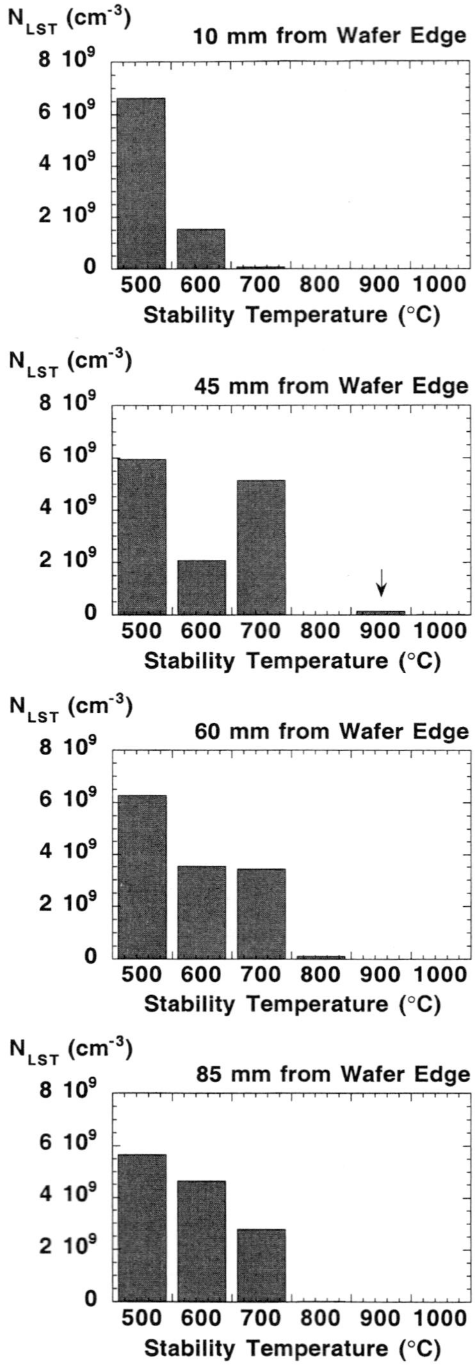

Fig. 1. Grown-in defect density spectra of characteristic regions of a 300 mm wafer with a transition from vacancy-rich to interstitial-rich. The arrow indicates the grown-in nuclei that are prone to develop a stacking fault ring during a subsequent wet thermal oxidation.

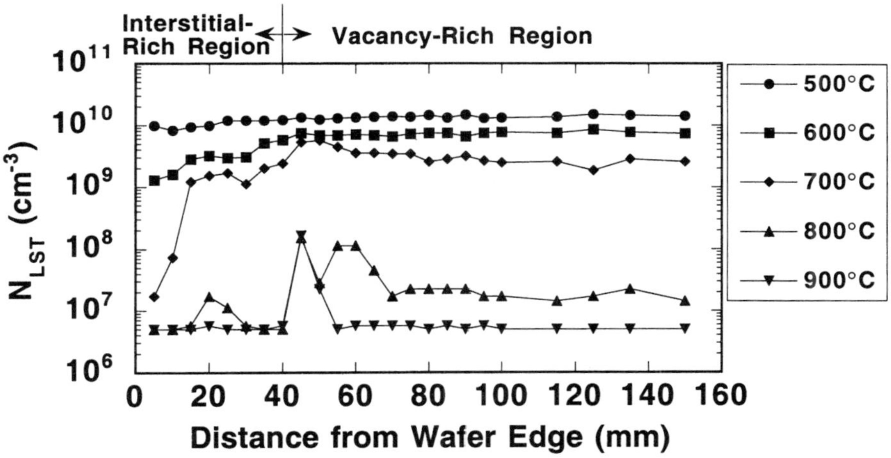

Fig. 2. Bulk defect density as a function of the radial position on the wafer shown for different starting temperatures of ramping with 1 K/min to 1000°C with a final anneal at 1000°C for 1 h in nitrogen.

which is slow enough that the growing nuclei always remain larger than the critical radius. The starting temperature of the ramp is also important because nuclei of lower size or stability temperature cannot be grown since their size is subcritical from the beginning. In order to grow all nuclei from a starting temperature, we use a ramp rate of 1 K/min.

Fig. 2 shows the radially dependent bulk defect density for different starting temperatures of ramping with 1 K/min up to 1000°C and final annealing for 1 h at 1000°C in nitrogen. The densities obtained for starting temperatures of 800 or 900°C are typical for processed wafers in a conventional CMOS technology. Only the larger nuclei can grow in the first thermal step of processing. This results in a very inhomogeneous bulk defect density across the wafer. However, with a ramp starting at 500°C, a radially uniform bulk defect density can be obtained across the whole 300 mm wafer. A bulk defect density of about 10^{10} cm^{-3} is the maximum obtainable density and ensures a high gettering efficiency even in these oxygen-lean wafers.

5. Conclusions

In summary, we have shown that:

1. In contrast to the size distribution of grown-in oxide precipitate nuclei, the total density of nuclei is nearly constant across wafers with a transition from vacancy-rich to interstitial-rich and independent of the dominant intrinsic point defect.
2. It is possible to obtain a laterally uniform and high density of gettering bulk defects in 300 mm wafers, even if they are obtained from a crystal with a transition from vacancy-rich to interstitial-rich. This can be achieved if the total density of grown-in oxide precipitate nuclei on each position of the wafer is grown by a slow ramp with 1 K/min starting from 500°C to 1000°C to form stable bulk defects.

Acknowledgements

The work has been supported by the Federal Ministry of Education and Research of Germany under contract No. 01 M 2973A. The authors alone are responsible for the contents.

The authors thank Mrs. B. Knoll and Mrs. S. Fröhlich for their technical assistance. Prof. A. Ourmazd and Dr. J. Marquardt are acknowledged for critical reading of the manuscript and helpful comments.

References

[1] M. Itsumi, H. Akiya, T. Ueki, M. Tomita, M. Yamawaki, The composition of octahedron structures that act as an origin of defects in thermal SiO_2 on Czochralski silicon, J. Appl. Phys. 78 (1995) 5984.

[2] G. Kissinger, G. Morgenstern, H. Richter, J. Vanhellemont, D. Gräf, U. Lambert, W. von Ammon, P. Wagner, D-defects and metal related flow patterns, Electrochem. Soc. Proc. 97–22 (1997) 32.

[3] J.G. Park, G.A. Rozgonyi, DRAM wafer qualification issues: Oxide integrity vs. D-defects, oxygen precipitates, and high temperature annealing, Solid State Phenom. 47–48 (1996) 327.

[4] P.J. Rocksnoer, M.M.B. Van Den Boom, Microdefects in a non-striated distribution in floating-zone silicon crystals, J. Cryst. Growth 53 (1981) 563.

[5] G. Kissinger, J. Vanhellemont, U. Lambert, E. Dornberger, R. Sorge, G. Morgenstern, T. Grabolla, D. Gräf, W. von Ammon, P. Wagner, H. Richter, Curriculum vitae of oxide precipitates: From nucleation during crystal growth to their final destination in processed wafers, Electrochem. Soc. Proc. 98–1 (1998) 1095.

[6] G. Kissinger, T. Grabolla, G. Morgenstern, H. Richter, D. Gräf, J. Vanhellemont, U. Lambert, W. von Ammon, Grown-in oxide precipitate nuclei in CZ silicon substrates and their role in device processing, Electrochem. Soc. Proc. 97–22 (1997) 74.

[7] G. Kissinger, D. Gräf, U. Lambert, T. Grabolla, H. Richter, Key influence of the thermal history on process-induced defects in Czochralski silicon wafers, Semicond. Sci. Technol. 12 (1997) 933.

[8] W. von Ammon, Expected limits for manufacturing very large silicon wafers, Solid State Phenom. 47–48 (1996) 97.

[9] G. Kissinger, G. Morgenstern, J. Vanhellemont, D. Gräf, U. Lambert, H. Richter, Internal oxidation of vacancy agglomerates in Czochralski silicon wafers during high temperature anneals, Appl. Phys. Lett. 72 (1998) 223.

[10] G. Kissinger, D. Gräf, U. Lambert, H. Richter, A method for studying the grown-in defect density spectra in Czochralski silicon wafers, J. Electrochem. Soc. 144 (1997) 1447.

[11] A. Borghesi, B. Pivac, A. Sassella, A. Stella, Oxygen precipitation in silicon, J. Appl. Phys. 77 (1995) 4169.

[12] E. Dornberger, W. von Ammon, D. Gräf, U. Lambert, A. Miller, H. Oelkrug, A. Ehlert, The impact of dwell time above 900°C during crystal growth on the gate oxide integrity of silicon wafers, Electrochem. Soc. Proc. 96–13 (1996) 140.

[13] K. Wada, N. Inoue, Oxide precipitate nucleation in Czochralski silicon, an insight from thermal donor formation kinetics, Electrochem. Soc. Proc. 86–4 (1986) 778.

[14] G. Kissinger, J. Vanhellemont, U. Lambert, D. Gräf, T. Grabolla, H. Richter, Conditions for the formation of ring-like distributed stacking faults in CZ-Si wafers, Jpn. J. Appl. Phys. Part 2 (Letters) 37 (1998) L306.

ELSEVIER

Microelectronic Engineering 45 (1999) 161–168

MICROELECTRONIC ENGINEERING

Effect of the structural state of the melt on the properties of silicon crystals

Anatolii Ya. Gubenko

Moscow Institute of Electronics and Mathematics, ul. Kuusinena 25-62, Moscow 125252, Russia

Abstract

Changes in the degree of supercooling and in the accompanying changes in the structure of the melt layer adjoining the growing silicon crystal was shown to result in changes in the properties of the crystal. Crystals of high quality were grown from the melt with a disordered structure. The formation of such a structure can be realized by introducing a special modifying impurity which initiates some phase transformations that lead to melt disordering.

Keywords: Silicon; Crystal growth; Local equilibrium; Phase transformation; Supercooling; Density

1. Introduction

When designing new technologies of crystal growth, the main attention is usually paid to growth regimes and the construction of the thermal unit (heater and thermal shields). The effect of the melt condition on the properties of crystal is usually ignored. At the same time, by varying the melt structure (condition), we may effectively control the quality of crystals grown [1].

The melt of even undoped silicon possesses dynamic microheterogeneities. Apart from statistically distributed atoms, any melt contains ordered atomic groupings (OAG) with a certain short-range order. The greater the size and the number of OAGs in the melt, the stronger the interatomic interactions in them. The OAGs have a fluctuational nature. Their dynamic lifetime is determined by the temperature and condition of the melt (its thermal prehistory). The energies of interaction between atoms existing in different groupings may differ significantly. However, they are correlated: an increase (decrease) in the energy of interaction between atoms in an OAG and the number of OAGs leads to corresponding changes in this energy when atoms of another nature are involved in them. The diversity of atomic groupings in the melt makes it possible to characterize its structure by the type and concentration of particles present. The melts in which the fraction of OAGs is small, possess a disordered structure with weakened interatomic interactions and a large fraction of statistically distributed atoms. The melts with a large fraction of OAGs have an ordered structure.

These dynamic microheterogeneities are very sensitive to external actions and growth conditions,

0167-9317/99/$ – see front matter
PII: S0167-9317(99)00129-X

which is especially noticeable in a layer of the melt adjacent to the solid–liquid interface (crystallization front). In essence, the crystal grows from this layer. The structural state of this layer also determines the processes occurring in the postcrystallization period in the crystal. Between the solid crystal and the conjugated layer of the melt, there is established a heterogeneous equilibrium even upon growing crystals at finite rates. In view of this equilibrium, any changes in the structure of the melt layer adjacent to the crystallization front lead to changes in the chemical potentials of the components in this layer and in the adjacent solid layer. The chemical potentials are related to the thermodynamic activities and, consequently, to the interatomic interactions. Therefore, their variation leads to changes in the entire set of properties in the conjugated layers of the solid and liquid phases. The presence of a local equilibrium has been confirmed by numerous experiments [1,2], as well as by the data that are considered below. This interrelation must necessarily be taken into account when growing crystals, especially ingots of large diameters.

2. Experimental

Silicon ingots 100 mm in diameter were grown in the $\langle 111 \rangle$ and $\langle 100 \rangle$ directions. The density of silicon in the solid and liquid states was determined by the gamma-ray method [3] with an accuracy of 0.05% (the accuracy of relative measurements was 0.01%). The rate of the temperature variation in a temperature range of 100–1600°C was 10 K/min; the density was measured every 30 s. In accord with the purpose of this work, the main attention was paid to changes in the density of the melt d'' and of the solid d' near the melting temperature T_m. Fragments of the density polytherms obtained are shown in Fig. 1. On samples cut from various (upper, middle, bottom) parts of ingots, we measured the lattice parameter (by the Bond method) and the concentration of charge carriers and determined the melting temperature and thermal effects (by the differential thermal analysis). The thermograms were constructed in the temperature difference (ΔT) vs. measurement temperature coordinates. The apparatus for thermal analysis was switched on through a computer and made it possible to control temperature with an accuracy of ± 1 K. The crucibles for measurements were made of fused quartz. The silicon ingots were doped with boron and phosphorus; in some cases, a second impurity was introduced to modify the structure of the melt and properties of crystals [2].

3. Results and discussion

As can be seen from Fig. 1, the values of d' and d'' near the melting point do not virtually differ. At higher temperatures, d'' sharply decreases jumpwise. At the same temperatures, the thermograms reveal an exothermic effect, which is superimposed on the endothermic effect due to the fusion heat (Fig. 2). Measurements of the steady-state vapor pressures (P) over silicon, germanium, pure metals, and their melts [4] showed that these pressures, just as d' and d'', at the interface between the solid and liquid phases at T_m were the same. At higher (lower) temperatures, P increases and then, after it reaches a maximum value, decreases. The oscillating character of the dependence of $1/P$, d' and d'' on the temperature is identical. It follows from these results that a local equilibrium is established at the interphase interface. The decrease in d'' and the increase in P with temperature can be interpreted as being due to a decrease in interatomic spacings in the melt and the development of a more disordered

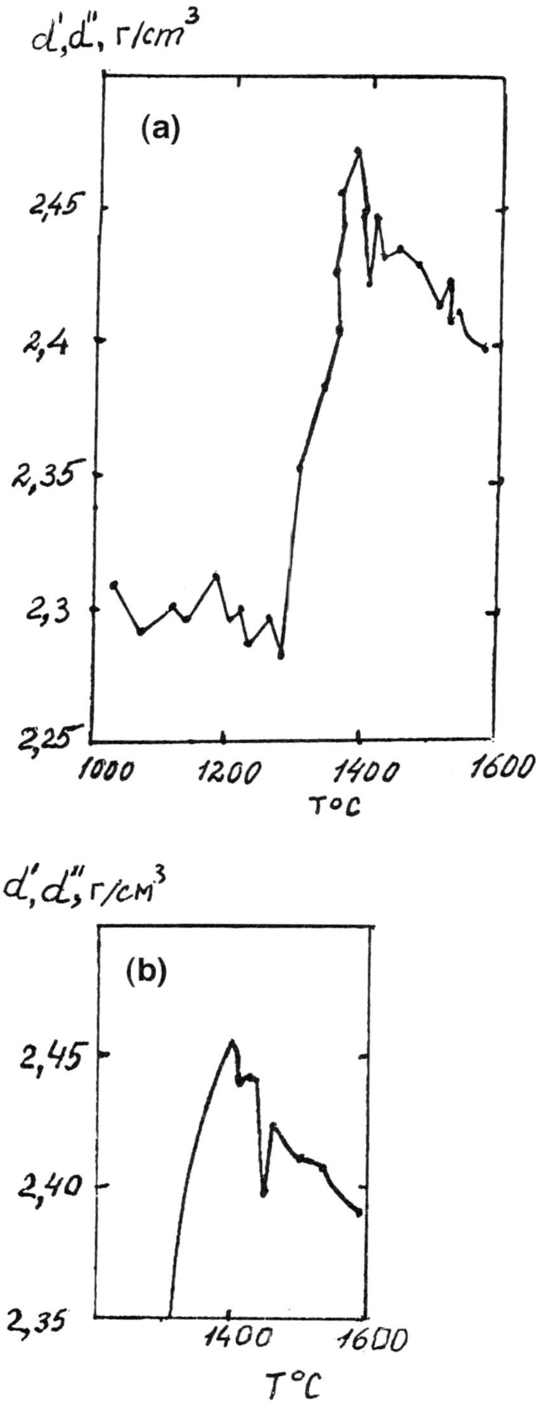

Fig. 1. Fragments of temperature dependences of density in solid and molten silicon doped with boron to a concentration of 5×10^{19} cm^{-3}. (a) Shows a magnified fragment of a polytherm for silicon doped with phosphorus to a concentration of 10^{19} cm^{-3}. Recorded in the regime of cooling. Key: 1. d, g/cm^3.

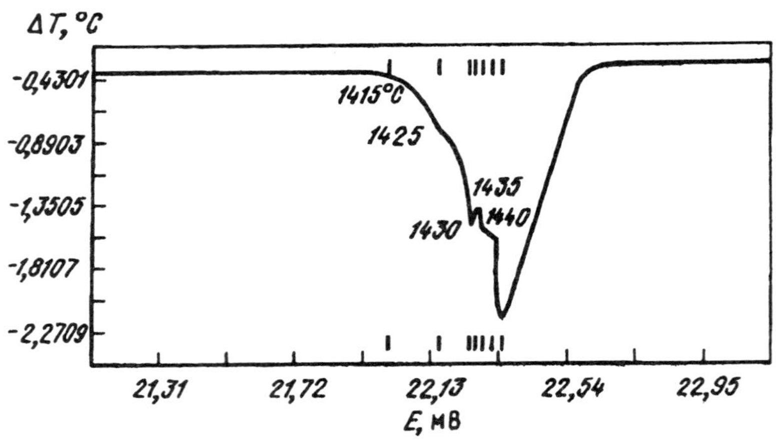

Fig. 2. Thermogram of silicon doped with boron to a concentration of 10^{19} cm^{-3} recorded in the regime of heating.

structure in the melt. The transition from one structure of the melt to another occurs non-monotonically; the temperature dependence of d'' exhibits oscillations, which are accompanied by thermal effects (Fig. 2). Similar changes occur prior to crystallization. In some cases, the thermal effects were fixed immediately before the liberation of the heat of crystallization (Fig. 3). Previous measurements of kinematic viscosity have shown that it varies after melting and before crystallization by sharp jumps.

From these findings and taking into account the proximity of the values of d'' and d' and the minimum of P at T_m, we suggested that the layer of the melt adjoining to the crystallization front was in an ordered state, which was supported by the force field of the surface atoms at the crystallization front. This force field tends to arrange the atoms in the adjoining layer of the melt in an ordered way, whereas their thermal motion exerts the opposite effect. As a result, there arises a liquid layer in which the interatomic interactions decrease with distance deeper in the melt until they achieve values characteristic of the bulk of the melt. The presence of such a specific layer of the melt and the effect of its structure on the properties of growing crystals is confirmed by a comparison of T_m of silicon dendrites and samples cut from single crystals. Both were grown from the melts with the same concentration of phosphorus and had close values of electrical resistivity (10–15 Ω cm). It turned out that the dendritic sample had $T_m = 1426$°C, whereas the single-crystal sample had $T_m = 1415$°C. After crystallization, the samples were repeatedly heated without removing the material from the crucible. The repeated measurements of T_m gave closer values.

In order to exclude the possible effect of the interaction of the silicon melt with the crucible material, we performed similar experiments using dendritic and single-crystal samples of germanium; in this case, the T_m of the dendritic sample was 4 K higher than that of the single-crystal sample. Only threefold repetition of the process of crystallization–melting of the sample that was dendritic in the initial state did ensure that its melting temperature became 937°C, as in the case of the single-crystal sample. The above differences in the melting temperature may only be due to the fact that the dendrite samples were crystallized from the melts that were supercooled to a much greater degree; in such melts the interatomic interactions near the solid–liquid interface are much stronger than when growing single crystals.

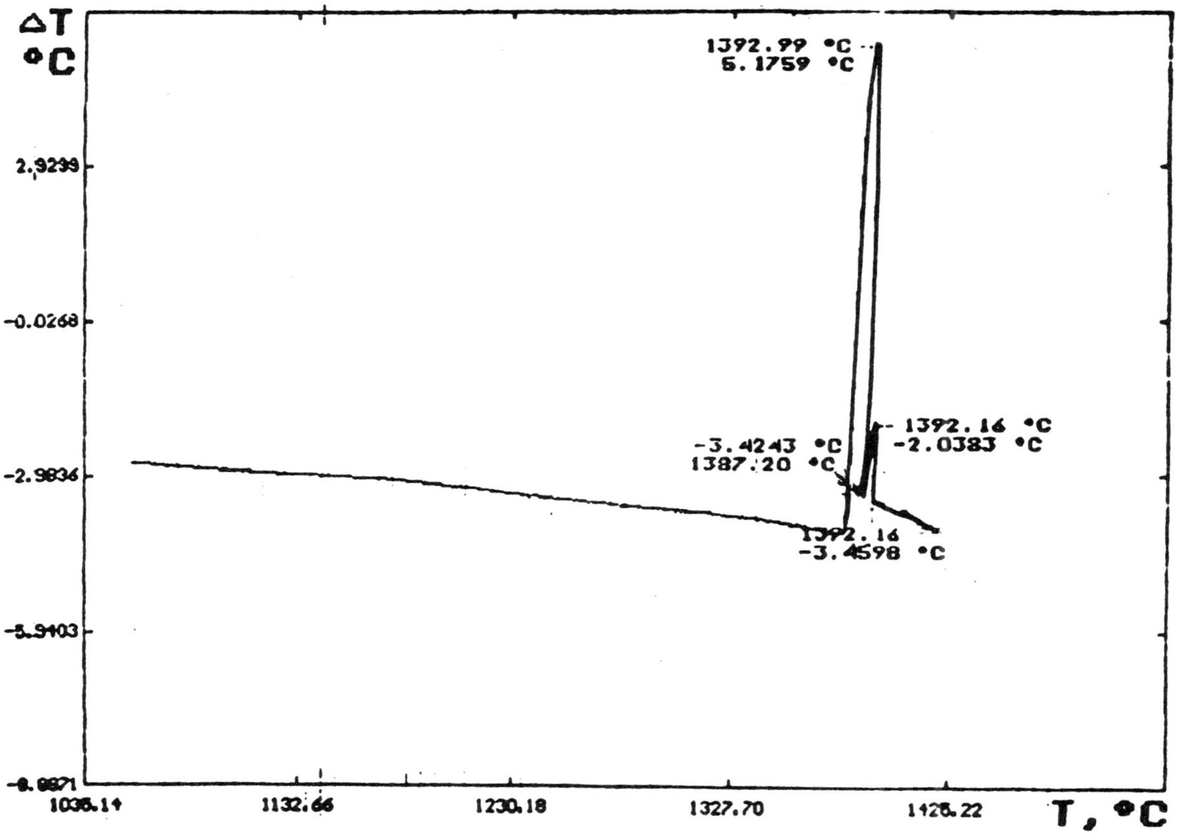

Fig. 3. Thermogram of silicon doped with phosphorus to a concentration of 4×10^{17} cm^{-3}.

Changes in the interatomic interactions depending on the extension of the transition layer in the melt near the solid–liquid interface were studied in the same melt using the results of the following experiment. The extension of this layer is also determined by the axial temperature gradient at the crystallization front (G). The greater the gradient, the narrower the layer and the smaller the differences in the interatomic interactions at its boundaries. The value of G, as was shown in our and other investigations, increases with increasing ingot length. This increase is caused by an increase in the crystal surface that loses heat through radiation and by the lowering of the position of the crystallization front, which moves deeper into the crucible as the crystal grows longer.

The character of interatomic interactions, as in the previous case, was estimated proceeding from the melting temperature of the samples that were cut from wafers located in the upper, middle, and lower parts of the ingots. We again measured the melting temperature T_m, the lattice parameter, and the concentration of charge carriers (holes), which were used to calculate the segregation coefficient of boron. The crystals were doped with boron to concentrations of 6×10^{18} to 1×10^{19} and 5×10^{14} to 1×10^{15} cm^{-3}. It was determined preliminarily that the T_m and the lattice parameters in such silicon are constant. The measurements showed that the T_m and lattice parameters of the upper and lower parts of the crystals differed by 5–6 K and 2–4×10^{-4} Å, respectively. The upper part of the crystal differs from the remainder in even higher concentrations of thermal donors (formed after annealing at 450°C), microdefects, and oxygen. The boron concentration distribution along the crystal length was steeper as compared with that calculated using Pfann's approximation and the boron segregation coefficient upon heavy doping proved to be more than unity in the lower parts of the crystals (Fig. 4). The variations of the electrical resistivity and of the lattice parameter along the crystal were symbatic. In the upper parts of the crystals, the changes were first insignificant but then became more pronounced. The observed changes in the measured values could only be due to 'concerted' changes in the interatomic interactions (component activities) in the conjugated layers of the crystal and melt being in local equilibrium and their variation in the course of growth.

From these findings and taking into account the proximity of the values of d' and d'' and the sharper decrease in d'' at higher temperatures, we may conclude that the state (structure) of the supercooled melt differs significantly from that of the superheated melt. The greater the supercooling and the larger the extension of the transition layer between the solid and the melt, the greater this difference. For this reason, the properties of crystals depend strongly on the degree of supercooling.

The structure and the extension of the transition layer at the solid–liquid interface is strongly affected by the intensity of melt stirring and, especially, by the growth rate. The lower the growth rate, the smaller the supercooling in this layer, and the smaller the degree of order in it. Correspondingly, the crystal obtained will differ in its properties from crystals grown at larger rates.

The above-noted effect of the area of the crystal surface on G and, consequently, on the structure of the melt layer adjacent to the crystallization front determines the established dependence of the properties of crystals on their diameter. It turned out that the smaller the diameter of a silicon crystal, the lower the density of microdefects of various types in it under other conditions being equal. This dependence is especially pronounced upon a comparison of the properties of crystals with diameters

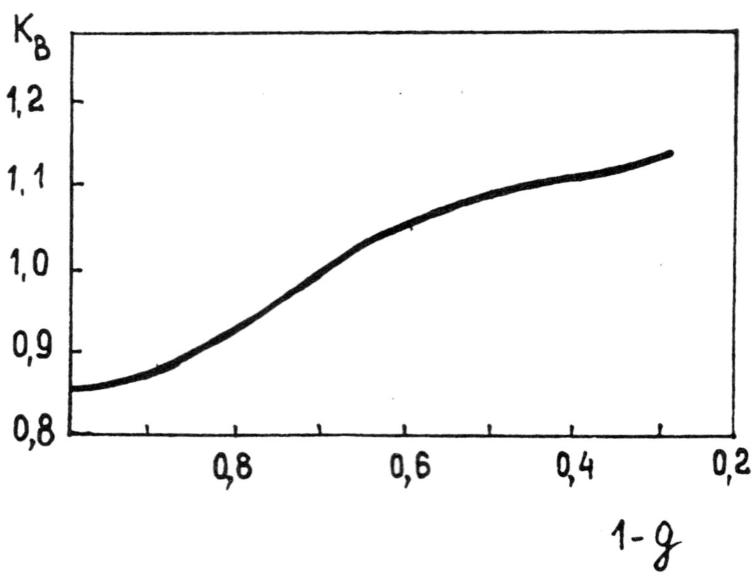

Fig. 4. Variation of the segregation coefficient of boron as a function of the fraction of the solidified silicon g.

of 40 and 100 mm. When growing crystals of 300 mm in diameter, this dependence must show itself up even more, taking into account the specific conditions of their growth. It was found that defect distribution in crystals of 300 and 100 mm differ significantly. For example, regions enriched in vacancies and interstitials were revealed in transverse sections [6]. According to [7], these regions can form due to structural inhomogeneity of the melt layer adjacent to the solid–liquid interface.

The degree of order (disorder) in the transition layer affects not only the electrophysical properties of crystals but also the growth kinetics, impurity distribution in the bulk of the crystals, and the perfection of their structure. During crystallization, the atoms on the growing surface, before they become built in the crystal lattice, make millions of unsuccessful attempts ('trials and errors'). The higher the atom mobility at the growth surface, the greater the number of such attempts. Taking into account the above-mentioned feature of the melts with a disordered structure, the atom mobility in such melts in the transition layer and at the growing surface is greatest in comparison with the melt regions with a close chemical composition but with an ordered structure. Therefore, upon crystallization from the disordered melts the probability for an atom to be built in energetically more favorable sites of a growing layer is higher. As a result, the atoms of the components of the system become more uniformly distributed relative to one another and the vacancies. Such crystals have narrower rocking curves as compared to the standard silicon, and the vacancies are distributed uniformly at the interphase interface and in the near-surface layers. Measurements of the lifetime and charge-carrier mobility have shown that their values in silicon crystals grown from the melts with a disordered structure are higher by 200–300 and 20–30%, respectively, than those observed in crystals grown from the melts with an ordered structure, with the other conditions being the same; they possess an enhanced thermal stability; they can be used without post-growth annealing; after annealing at temperatures corresponding to technological schedules used upon device manufacture, their resistivity changed only slightly; the processes of the formation of thermal donors and other complexes. The above advantages were independent of the growth direction of the crystals ($\langle 111 \rangle$ or $\langle 100 \rangle$).

Taking into account the specific features of the process of growing silicon single crystals, the possibilities for the formation of melts with a desired structure are limited. We developed a new method for producing melts with a disordered structure, which is based on a new phenomenon related to the impurity effect on the initiation of specific phase transformations occurring in a certain concentration range [1,2]. This range separates the parent phase from the new phase with a smaller free energy under given parameters of state. In this range of impurity concentrations, silicon and other materials search the new state in the 'trial-and-error' manner through large-scale fluctuations. Therefore, in this interval of concentrations, there arise oscillations of various properties. The oscillations consist of a series of equilibrium microstates that arise through large-scale fluctuations. In the melt, the degree of disordering increases along one of the branches of such oscillations and decreases along the other branch.

The impurities commonly used for doping silicon initiates, as a rule, only one phase transformation in the concentration ranges applied in practice. In order to apply this effect to silicon irrespective of the doping level, we introduced a second impurity, which initiated the desired phase transformation. The concentration of this impurity must correspond to that at which a disordered structure is formed in the melt. In view of the development of local equilibrium at the interphase interface, a new mechanism of impurity influence on the properties of the melts and semiconductor solid solutions formed from these melts manifests itself in this range [5]. For this reason and taking into account the fluctuational nature of this phenomenon, this mechanism of searching a new phase was called the equilibrium self-organization.

The choice of a second impurity, which was called modifying impurity, depends on the concrete problem. In particular, this choice can be determined by the doping level, the need of using internal gettering, etc. To produce the melt with a disordered structure, some isovalent impurities and rare-earth metals were used.

The paradox of this phenomenon consists in the fact that the presence of foreign atoms of a second impurity in the crystal lattice of crystals grown from the melts with a disordered structure does not impart but, on the contrary, improves the quality of silicon. In such crystals, the lifetime and mobility of charge carriers are higher in comparison with crystals doped with a single impurity, and optical and some other properties are also improved.

Based on this method of modifying melt properties, several techniques of growing high-quality silicon crystals were developed. Their application for the production of silicon crystals 300 mm in diameter gave a positive effect.

References

[1] A.Ya. Gubenko, Rasplavy (1) (1991) 27–33.
[2] A.Ya. Gubenko, Tsvetn. Met. (7) (1995) 44–49.
[3] O.B. Borovskii, in: Gamma-metod v metallurgicheskoi promyshlennosti (Gamma Method in Metallurgy), Novosibirsk, 1981, p. 22.
[4] Y.A. Priselkov, A.V. Tseplyaeva, S.A. Zaitseva et al., Zh. Fiz. Khim. 42 (3) (1968) 772.
[5] A.Ya. Gubenko, Izv. Vyssh. Uchebn. Zaved., Mater. Elektron. Tekh (1) (1998) 20.
[6] G. Kissinger, J. Vanhellemont, U. Lambert, in: Book of Abstracts E-MRS'1998, pp. F–4, June.
[7] A.Yu. Gubenko, Tsvetn. Met. (5) (1986) 59–62.

ELSEVIER

Microelectronic Engineering 45 (1999) 169–182

MICROELECTRONIC ENGINEERING

Characterization of 300 mm silicon-polished and EPI wafers

Steven Shih[a,*], Chi Au[a], Zach Yang[a], Troy Messina[a], Randal K. Goodall[a], Howard R. Huff[b]

[a]*International 300 mm Initiate, 2706 Montopolis Drive, Austin, TX 78741, USA*
[b]*SEMATECH, 2706 Montopolis Drive, Austin, TX 78741, USA*

Abstract

Maturity of 300 mm polished wafers and early epi wafers were evaluated in respects of particles, flatness, metal contamination, and epitaxy thickness. Data of 300 mm polished wafers showed encouraging characteristics comparable to state-of-the-art 200 mm prime wafers. Preliminary characterization of 300 mm epi wafers revealed that dominant localized light scatterers (LLS) with sizes more than 1 μm were epitaxy growth-related surface or bulk imperfections.

Keywords: Evaluation of wafer maturity; 300 mm silicon wafers; Characterization of silicon wafers

1. Introduction

The mission of International 300 mm Initiative (I300I) is to ensure 300 mm wafer processing capability available to the silicon industry at the right time. The focus is to make ready of 300 mm 0.25 μm process equipment in 1997 and initiate the demonstration of 0.18 μm capable equipment in 1998–99. Since the cost of 300 mm silicon wafers is projected to be as much as 30% of the total wafer processing cost [1], it is essential to align all the silicon wafer suppliers to be more cost effective in order to meet the forecasted 300 mm production ramp-up in 1999–2000. LLS, metal contents and flatness of incoming wafers have been studied at I300I to evaluate the quality of potential silicon suppliers [2]. The preliminary characterization of developmental epi wafers, which are sampled in the first half of 1998, provides a snapshot of the performance of these early available 300 mm epi wafers.

2. Particles and crystal-originated pits (COPs) separation

1997 National Technology Roadmap for Semiconductors (NTRS) [3] modeled the killer defect size as half of critical dimension (CD) for a given technology. LLS can be reliably monitored by modern

*Corresponding author. Tel.: + 1-5122-356-7044; + 1-512-356-7848.

0167-9317/99/$ – see front matter
PII: S0167-9317(99)00134-3

surface inspection systems (such as equipment P1 used at I300I) to as low as 0.08 μm on bare silicon wafers. Large LLS with sizes ≥0.2 μm were regularly found in early developmental polished wafers from silicon suppliers. The presence of excessive COPs in early 300 mm wafers accounts for a significant portion of observed LLS smaller than 0.14 μm. Presently defects inspection industry faces at least three primary road blocks: (1) separation of real particles from COPs, (2) reduction of small particles and COPs by optimized control or/and novel processes and (3) global size standardization of polystyrene latex (PSL) spheres among different suppliers.

Based on the methodology originally developed by Wacker [4], real particles and COPs can be readily separated from each other. Experiments performed by Wacker applied a series of scan/cleaning/scan to distinguish stationary COPs from removable particles. The capability of separating real particles from COPs is critical to applications in wafer cleaning, defects reduction as well as advanced cleanroom technology.

Surfscans were performed with both wide and narrow channels of P1 turned on to collect LLS events between 110 nm and 1000 nm. The data were stored as ".tff" files and converted to text files using "*Textout*" software. Interactive Data Language (IDL version 5.0.2) was then used to separate double-counted events, which were logged by both wide and narrow channels, from individual events. The resulting double-counted events were further analyzed to extract the size delta between channels.

There is a strong belief that due to the physical shape and dimensional difference, reflected light from COPs is preferentially scattered into narrow channel detector of P1 and are frequently reported as a larger LLS compared to the size appeared in wide channel. By gaussian fitting the histogram of size delta, as shown in Fig. 1, two distinguished groups of events with one group's peak close to zero

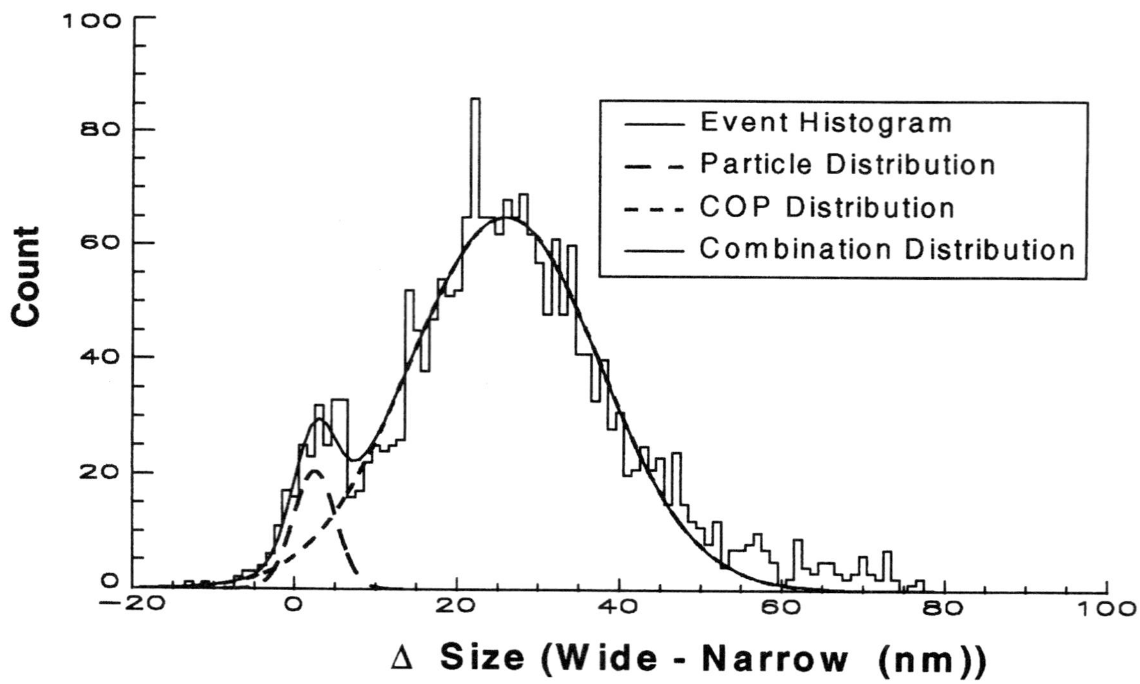

Fig. 1. Event size on as-received 300 mm wafers ≥ 100 nm.

delta and the other group's peak at ~25 nm delta can be obtained. The group with the peak close to zero delta is presumably composed of real particles. On the other hand, the group which shows a clear size delta between wide and narrow channels is attributed to COPs.

Several limitations apply to the approach described above. First of all, only LLS detected in both wide and narrow channels can be separated by the guassian fitting. Individual events which are only detected by either one of the channels can not be classified by size delta approach. Secondly, the success rate of catching COPs and flawlessly distinguishing COPs from particles is still unknow. The development of the size delta approach is still in "belief" or model simulation phase and lack of direct support from systematic experiments. Third, for a fine particle-grade wafer the total LLS count for sizes ≥0.12 μm would be likely less than 100. For these superior particle wafers, how well the gussian distribution can fit into the histogram needs to be addressed. More background works need to be done before the size delta approach can be widely accepted.

3. 300 mm polished wafers characterization

3.1. Polished wafers LLS

The LLS results of five silicon suppliers are listed in Table 1. The minimum LLS size has been pushed from 0.12 μm in 1997 toward 0.09 μm in 1998 as dictated by 0.18 μm process requirement. Compared to data collected in 1997, information in Table 1 shows remarkable improvement of LLS@0.12 μm from originally more than 100 to close to 50 this year. Supplier C even shows a record lot average of 14 LLS@0.12 μm, which is well below the average ~50 of the rest of suppliers. This

Table 1
LLS summary of 300 mm as received polished wafers

Q1-Q2 of 1998 bin size (μm)	Composite DFW 0.09–0.20, DFN 0.2–1.0 Supplier (wafer type)						
	A(furnace)	F(furnace)	A(particle)	C(particle)	D(particle)	E(particle)	A(litho)
0.090–0.100	282	1367	278	47	280	117	329
0.100–0.120	209	1726	190	51	85	67	184
0.120–0.140	21	694	24	5	6	10	18
0.140–0.160	12	571	8	2	4	7	6
0.160–0.180	8	282	4	1	2	4	3
0.180–0.200	2	45	2	0	1	2	1
0.200–0.300	32	85	9	3	17	10	15
0.300–0.500	12	23	4	2	9	5	5
0.500–1.000	8	35	4	1	4	6	11
Total >0.090	586	4827	522	111	408	228	573
Total >0.120	95	1735	55	14	44	44	60
Total >0.160	63	470	23	7	34	27	35
>0.090/>0.160	33.9	10.4	29.0	24.6	19.0	9.9	20.8
DFW avg haze (ppb)	248	83	254	26	59	23	234

great enhancement indicates excellent progress made in ingot pulling control for COPs suppression since last year.

Most of silicon suppliers used either slow crystal pulling (<0.1 mm/min) or heavily doped p-type substrate last year for COPs reduction. The average observed LLS ≥ 0.09 μm in the first half of 1998 shown in Table 1 is ~300. The desired target of allowed LLS is originally set to 100@0.09 μm at I300I, however, for cost control reason the target was relaxed to 200/wafer. Even so, most of the supplier today would still fall out of the I300I's specifications. Down to 0.09 μm LLS, a more aggressive approach is needed to further reduce the COPs amount. H_2 anneal was previously proposed to be one of the options for this purpose.

Control of haze level is another subject critical to the reduction of LLS@0.09 μm and below. Due to the promoted sensitivity of current metrology tools, high haze level wafers occasionally can cause false counts of small LLS as well as mediocre data repeatibility. ~50 ppb level of haze is recommended as maximum level haze to prevent above issues based on I300I's past experience. Some suppliers, such as supplier A, have an elevated haze level and therefore more liable to the noise-related issues.

3.2. Trace Metal Contamination and Wet Bench Capability

Trace metals of incoming 300mm wafers was monitored by vapor phase decomposition (VPD) inductively coupled plasma mass spectrometry (ICP/MS). Surface contaminants and native oxide were dissolved by HF vapor and then collected by the VPD system. The HF droplet was diluted in 1% HNO_3 to about 1 cm^3 before it was fed into the ICP/MS system for analysis. NTRS's trace metals recommendations are adopted as guidelines for metal contamination at I300I. Some metals such as calcium and sodium are relatively evironmental-rich and therefore highly sensitive to ambient control during the data collection.

Both as-received wafers and the wafers cleaned by in-house wet benches at I300I were evaluated (see Tables 2 and 3). The sodium and calcium levels are artificially high primarily due to manual operation during experiments. Both suppliers B and D demonstrate acceptable levels of metal contents, while the rest of supplier are not far from the tolerance limits.

Table 2
Silicon suppliers incoming wafer metal capability ($\times 10^{10}/cm^2$) (supplier/test date)

Element	NTRS Spec	B (11/97)	D (12/97)	C (12/97)	E (12/97)	F (12/97)
Na	≤ 2.5	2.9	16	64	450	2800
Al	≤ 10	5.4	5.3	7.9	16	88
K	–	1.1	0.8	5.7	18	190
Ca	≤ 2.5	17	37	17000	9400	86000
Cr	≤ 2.5	0.013	0.075	0.65	0.32	0.39
Fe	≤ 2.5	1.4	1.8	15	9	71
Ni	≤ 2.5	0.046	0.047	7.9	3.6	39
Cu	≤ 2.5	0.092	1.6	10	1.9	6.3
Zn	≤ 10	9.2	3.3	53	120	560

Table 3
Trace metal level of wet cleaned wafers ($\times 10^{10}/cm^2$) (wet Bench/test date)

	SIA Spec	W1 (5/98)	W2 (2/98)	W4 (12/97)	W5 (12/97)	W3 (11/97)
Na	< 2.5E 10	6.08E + 10	5.73E + 10	2.28E + 10	7.50E + 09	2.77E + 11
K	–	1.61E + 10	1.06E + 10	6.76E + 10	< 7.2 E + 08	2.11E + 10
Ca	< 2.5E 10	7.97E + 11	3.68E + 11	6.10E + 11	2.25E + 11	4.38E + 11
Al	< 1.0E 11	6.03E + 10	3.73E + 10	3.01E + 10	3.00E + 10	5.10E + 10
Cr	< 2.5E 10	< 3.3 E + 07	2.29E + 09	2.70E + 09	4.80E + 07	4.84E + 08
Fe	< 2.5E 10	8.3 E + 09	1.98E + 10	1.62E + 10	1.46E + 10	1.31E + 10
Ni	< 2.5E 10	1.12E + 10	5.8 E + 08	1.03E + 10	1.75E + 10	3.43E + 09
Cu	< 2.5E 10	7.08E + 09	1.73E + 10	2.13E + 10	1.80E + 10	1.27E + 09
Zn	< 1.0E 11	3.37E + 10	4.88E + 10	2.55E + 10	1.30E + 10	2.17E + 10

Out of SIA Spec is shaded
Lowest in each kind is inverted in color

Wet benches W1 to W4 are equipped with conventional batch-type cleaning stations, whereas W3 is single-wafer spin-on type compact cleaning tool. As indicated in Table 3, the conventional wet bench W5 has the lowest alkaline level detected, while the single wafer tool W3 shows almost comparable cleaning capability to the conventional ones.

3.3. Polished wafers flatness

As photolithography evolves, wafer surface flatness measured by GBIR, GFLR, and SFQR [6] have become more important than ever. The newly introduced scanning steppers will even require the measurement of scanning site flatness parameter, SFSR [7,8] in the near future.

SFQR was measured for polished wafers of four suppliers by equipment F1, as shown in Fig. 2. Equipment F1 employs capacitative probe to detect the surface topography change with a relatively large probe size of 4 mm × 4 mm. All the silicon suppliers except supplier D can reach a pretty high standard of more than 90% of measured sites having a SFQR smaller than 0.18 μm. From the past history on supplier D we learn that the F1 data on supplier D could be impacted by a so-called "donor kill" issue [5], which is associated with the relative insensitivity of some areas on the highly doped p-type substrate to the capacitative probe. Fig. 3 further shows a typical cumulative SFQR histogram of five wafers from supplier D. The majority of the sites on the surface are peaked at 0.04 μm, leaving a wide process margin for future lithography improvement.

Equipment F2 applies a much smaller optical probe size 12 μm × 12 μm for flatness measurement, which enables 1 mm edge exclusion as opposed to 3 mm exclusion for equipment F1. The initial correlation between these two techniques was studied for SFQR and wafer thickness. As expected from their different working principles and probe sizes, SFQR data did not correlate well between F1 and F2, as shown in Fig. 4(a). However, the wafer thickness and other parameters such as SBIR are shown in Fig. 4(b) and (c) to have good agreement between these two different measurement methods.

Table 4 summarizes three suppliers' flatness data except SFQR. Two suppliers have pretty good control margin to current I300I's thickness, warpage, and GBIR specifications. Only supplier D exhibits marginal warp and GBIR data, which might require some improvement in the future.

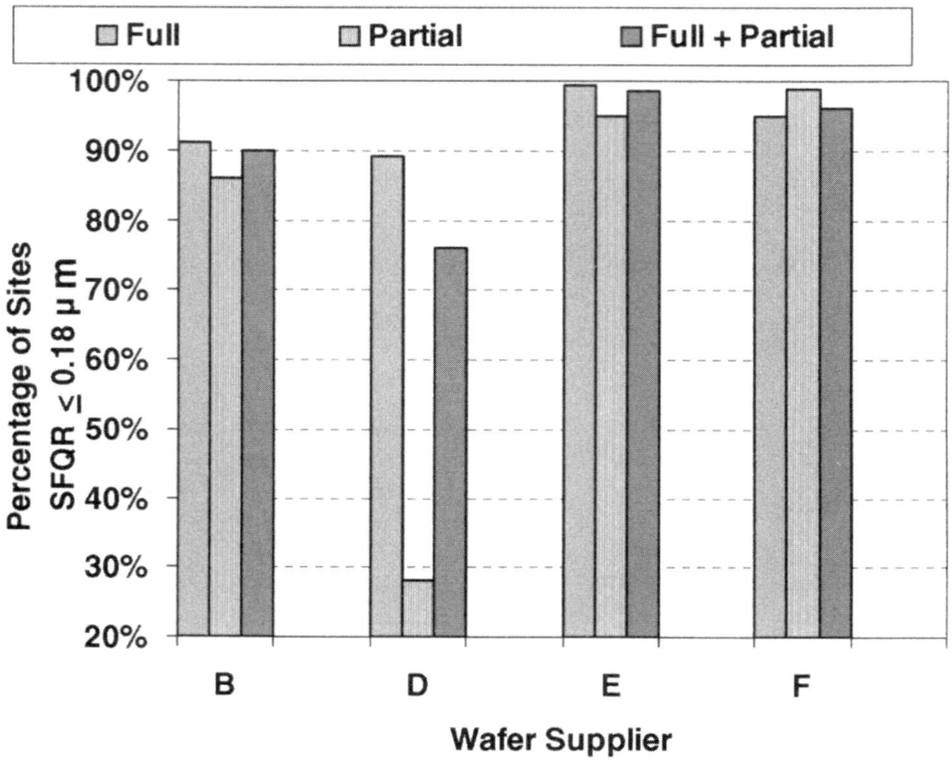

Fig. 2. Summary of SFQR of incoming polished wafers.

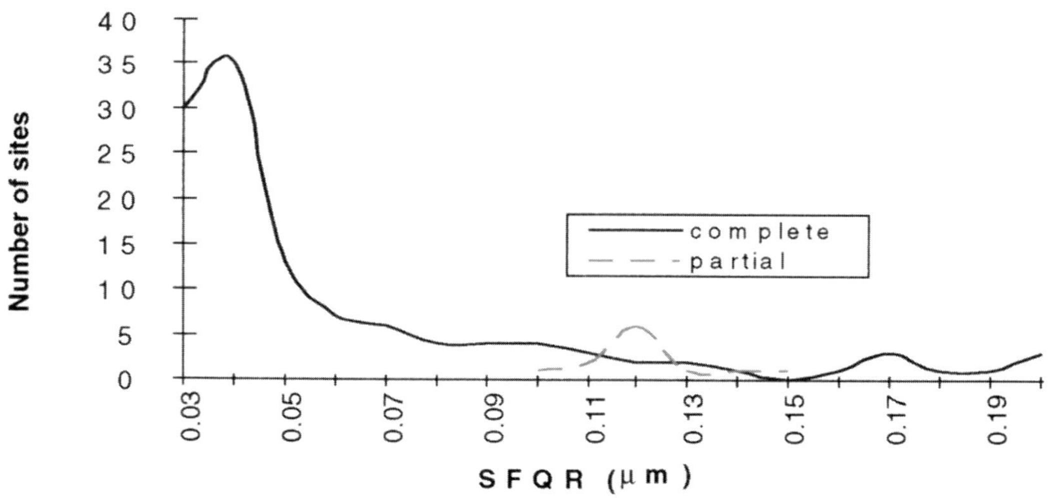

Fig. 3. Cumulative SFQR histogram from supplier D.

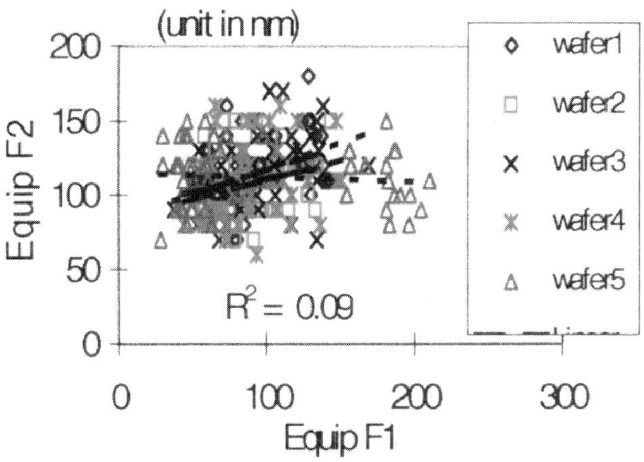

(a) SFQR - Supplier D

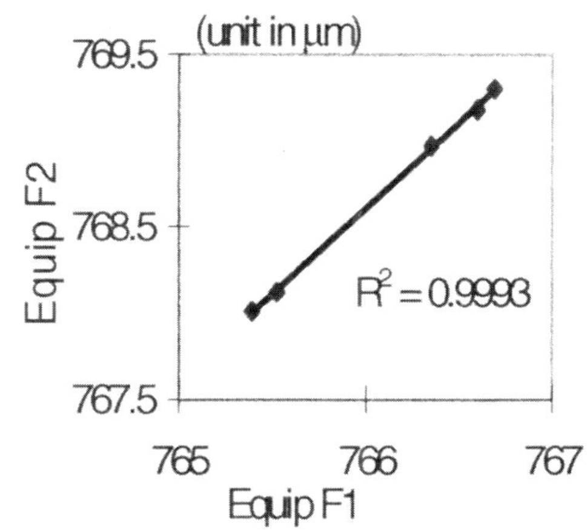

(b) Average Thickness - Supplier D

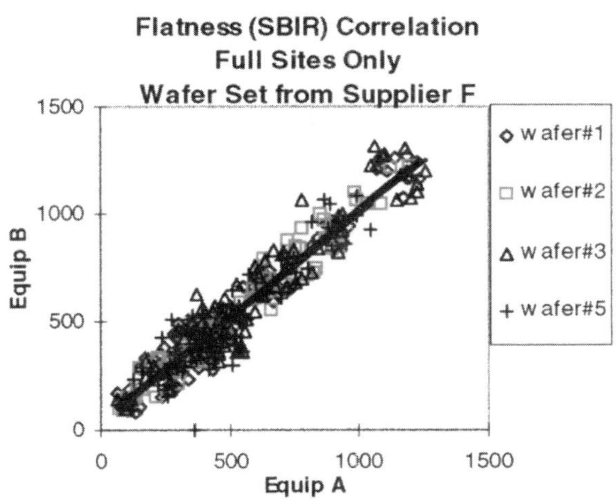

(c) SBIR correlation on wafers from supplier F

Fig. 4. Correlation results between flatness measurement tools (Equipment F1 and F2).

4. 300 mm Epi wafers characterization

Two types of 300 mm epi wafers from supplier B, C, and D are available and checked out by I300I for early process development. The major difference between them is the concentration of substrate doping, namely heavily doped p^+ substrate with resistivity ~10 mΩ-cm from supplier C, D and lightly doped p^- substrate with resistivity ~10 Ω-cm from supplier B. Supplier B, D use epi growth chamber manufactured by AMAT, whereas supplier C uses reactors made by ASM. Preliminary data collected by I300I silicon team provide insights into the maturity of these developmental epi wafers.

Table 4
Flatness data by equipment F1 (site size 25 mm × 25 mm, 3 mm Edge Exclusion, data in μm)

Supplier	Wafer type I300I spec	Ave Thk 775±25	Warp ≤ 50	Bow –	GBIR ≤ 3	GFLR	SBIR	SFLR
B	litho (for .25 μm)	776.67	29.86	13.42	0.57	0.62	0.45	0.46
B	litho (for .25 μm)	777.14	26.66	13.94	0.72	0.72	0.55	0.55
B	litho (for .25 μm)	777.03	26.76	12.17	0.64	0.55	0.42	0.43
B	litho (for .25 μm)	777.02	28.68	14.21	0.74	0.58	0.46	0.46
B	litho (for .25 μm)	777.01	27.42	8.83	0.89	0.81	0.61	0.59
B	litho (for .25 μm)	776.94	37.76	23.40	1.10	0.77	0.61	0.65
	Mean	776.97	29.52	14.33	0.78	0.67	0.52	0.52
	Stdev	0.16	4.22	4.86	0.19	0.11	0.08	0.09
D	litho (for .18 μm)	766.76	43.32	22.20	2.29	2.44	0.97	0.99
D	litho (for .18 μm)	766.10	43.55	23.66	2.67	2.42	1.02	0.99
D	litho (for .18 μm)	768.53	40.21	19.30	2.49	2.44	1.05	1.04
D	litho (for .18 μm)	765.60	40.36	21.17	2.54	2.51	0.99	1.00
D	litho (for .18 μm)	770.91	48.50	30.94	2.20	2.21	1.13	1.13
D	litho (for .18 μm)	770.47	40.14	24.69	2.48	2.53	1.22	1.27
	Mean	768.06	42.68	23.66	2.45	2.43	1.06	1.07
	Stdev	2.27	3.25	4.03	0.17	0.11	0.09	0.11
E	litho (for .18 μm)	783.00	19.10	5.27	1.04	1.16	0.70	0.72
E	litho (for .18 μm)	777.03	19.84	2.34	0.43	0.45	0.31	0.32
E	litho (for .18 μm)	776.44	22.66	− 1.24	0.46	0.46	0.30	3.04
E	litho (for .18 μm)	774.35	23.29	− 0.99	0.45	0.47	0.32	0.32
E	litho (for .18 μm)	779.87	21.36	4.18	0.89	0.98	0.62	0.64
	Mean	778.14	21.25	1.91	0.65	0.70	0.45	1.01
	Stdev	3.36	1.79	2.96	0.29	0.34	0.19	1.15

4.1. Epi thickness profile and inter-tool Epi thickness correlation

The most widely used techniques for epi thickness measurement are based on optical interference of infrared reflectance, such as Fourier-transformed infrared spectroscopy (FT–IR). Because of non-destructive and high throughput nature, FT–IR becomes a standard tool for epi thickness monitor in the industry. However, for supplier B whose lightly doped substrate is relatively transparent to infrared, destructive methods such as secondary ion mass spectroscopy (SIMS) or spreading resistace measurement (SRP) would be more appropriate. As shown in Fig. 5, epi thickness of the wafers grown by supplier C and D using different type of reaction chambers has a similar across wafer profile. Epi wafers from both suppliers exhibit symmetrical thickness waviness with variation ~0.15 μm. Supplier C has a significant roll-off close to 20 μm from the wafer edges and a steeper dip ~0.1 μm in depth at the wafer center. These film thickness features would certainly pose a serious threat to more and more stringent flatness requirement set by the advanced photolithography.

The correlation among different IR-based thickness measurement tools is illustrated in the boxplot of Fig. 6. Tools A is a conventional FT–IR tool using interferogram sideburst position for determining epi thickness. Tool B applies a model-based fitting of reflectance spectrum and is capable of extracting epitaxial transition regime as well as carrier doping profile in addition to epi thickness. Tool C is an advanced ellipsometer which uses polorization change of light to determine the refractive

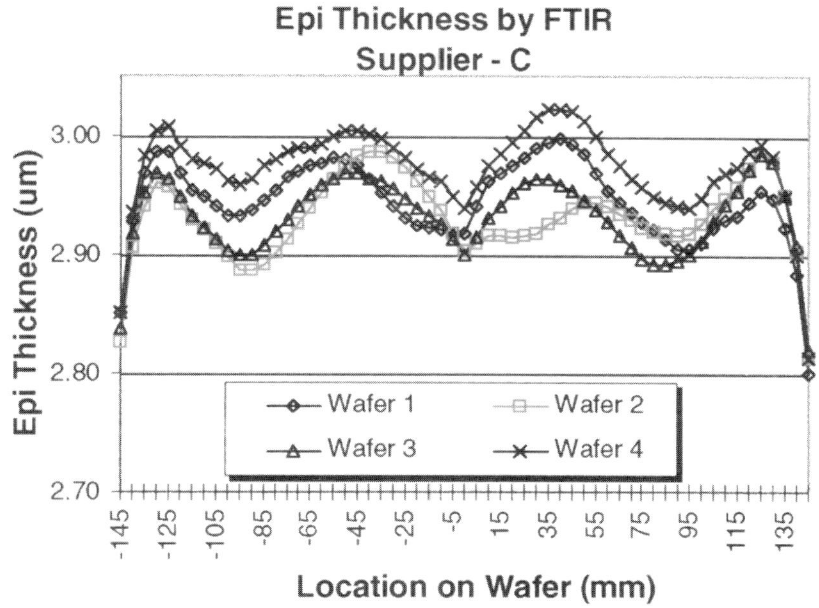

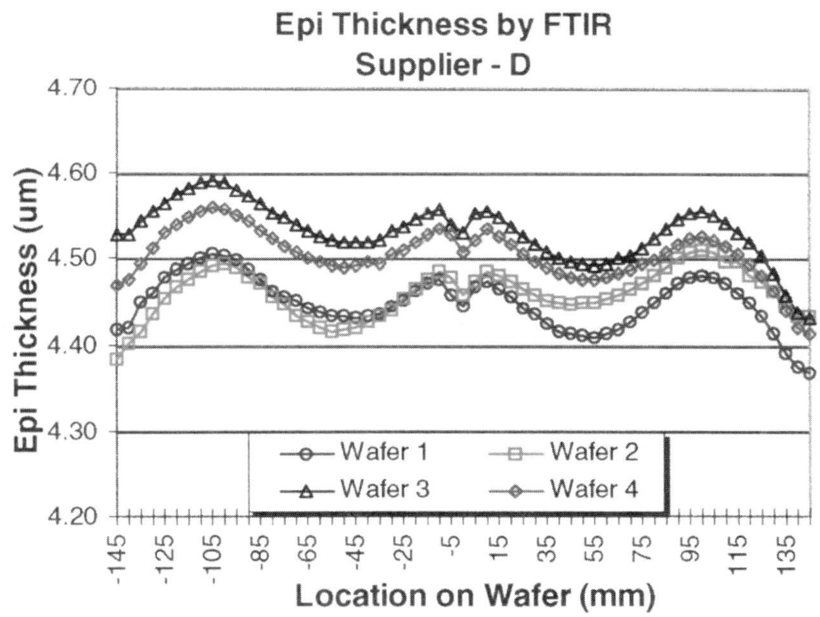

Fig. 5. Epi thickness profiles by conventional FT–IR for suppliers C and D.

index and the film thickness. As observed in Fig. 6(a) and (b), reading of tool A is consistently higher than tool B by ~0.02 μm, while reading of tool C is slightly lower than tool B and the difference depends on the wafer characteristics. These observed thickness differences are most likely determined

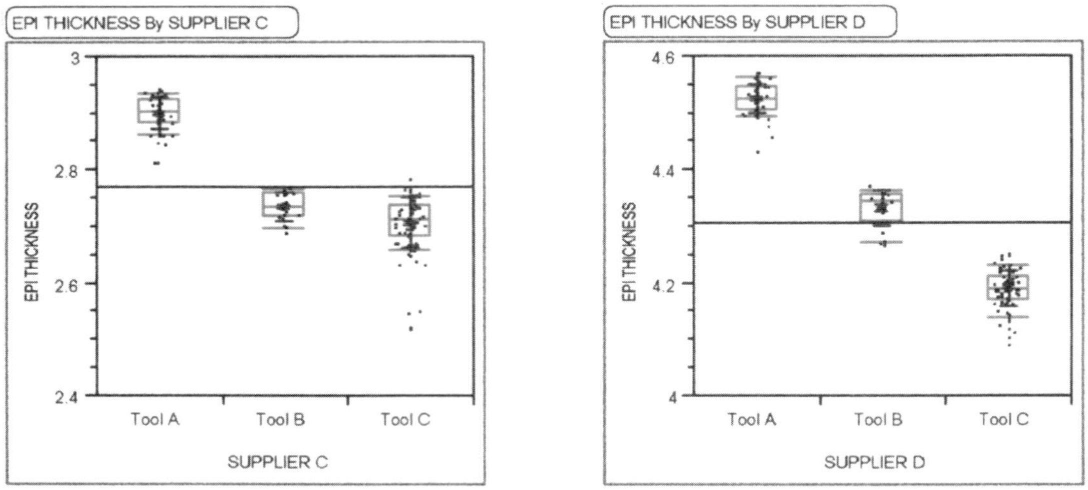

Means and Std Deviations

Level	Number	Mean	Std Dev	Std Err Mean
Tool A	59	2.90164	0.029926	0.00390
Tool B	17	2.73806	0.025511	0.00619
Tool C	113	2.70737	0.046281	0.00435

Means and Std Deviations

Level	Number	Mean	Std Dev	Std Err Mean
Tool A	59	4.52497	0.028055	0.00365
Tool B	17	4.33271	0.033615	0.00815
Tool C	113	4.19012	0.033006	0.00310

Fig. 6. Epi thickness comparison between tools.

by how the tools respond to the epitaxial transition zone. By model-based fitting, tool B theoretically has a chance to resolve the transition regions better than other tools. SIMS and SRP are currently in progress to confirm the above argument.

4.2. *Epi flatness*

SFQR site flatness of epi wafers was shown in Fig. 7. In spite of epi thickness waviness, the site flatness of supplier C is well controlled to be within the current 0.18 μm specification, consistent with the measurement made on their polished wafers. In addition, for the first time we observe that partial sites flatness has better control than full site flatness. The rest of suppliers has the flatness data comparable to their polished wafers.

Table 5 summaries the rest of flatness-related parameters for epi wafers. Notice that the warpage and GBIR of epi wafers from supplier D fall out of current I300I specifications, which is consistent with marginal data shown in their polished wafers.

4.3. *Epi LLS*

Table 6 listed the LLS performance measured by equipment P1 at I300I. In addition to developmental epi wafers, supplier C provides their early production epi wafers for the initial characterization. The impressed result of less than 65 LLS @0.09 μm from these production epi wafers signal the rapid learning of some silicon suppliers over the past 6 months. For majority of

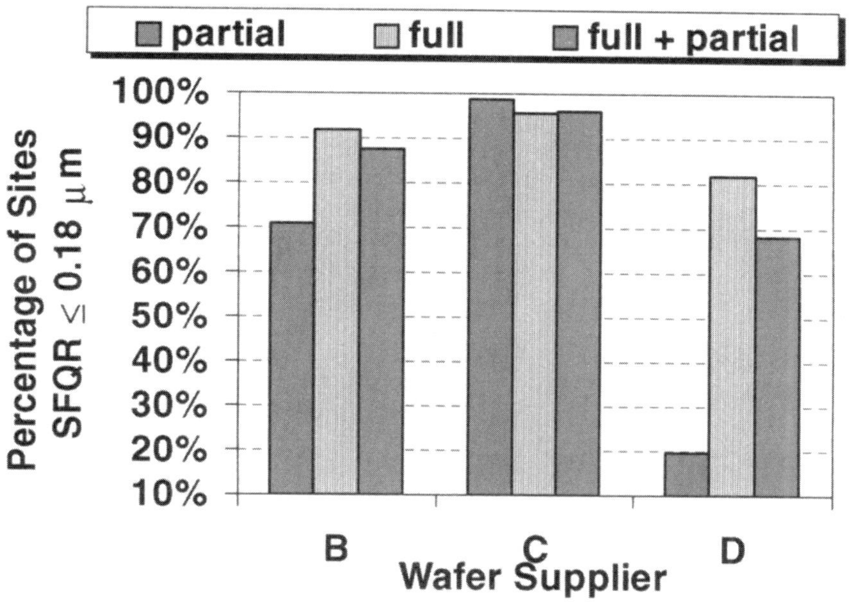

Fig. 7. SFQR site flatness for epi wafers.

suppliers, growth of epitaxy layer significantly reduce small LLS with sizes ≤ 0.18 µm. This LLS reduction was primarily attributed to the coverage of COPs by the grown silicon film. Oppositely, LLS with sizes ≥ 0.18 µm remarkably increases beyond the level of normal polished wafers in these early epi wafers. Additionally, increase of haze level from ~30 ppb in polished wafers to ~100 ppb

Table 5
Flatness data summary for epi wafers

Supplier	Wafer type I300I spec	Ave Thk	Warp ≤50	Bow –	GBIR ≤3	GFLR	SBIR	SFLR
B	TW305	780.92	12.63	−0.52	0.74	0.63	0.40	0.39
B	TW305	781.03	14.24	−0.17	0.68	0.56	0.38	0.40
B	TW305	780.86	11.40	1.62	0.57	0.50	0.38	0.38
	Mean	780.94	12.76	0.31	0.66	0.56	0.39	0.39
	Stdev	0.09	1.42	1.15	0.09	0.07	0.01	0.01
C	TW305	774.79	35.36	8.52	2.82	2.71	1.40	1.37
C	TW305	774.71	36.47	7.10	2.72	2.69	1.16	1.17
C	TW305	774.06	35.74	7.22	2.70	2.65	1.39	1.40
	Mean	774.52	35.86	7.61	2.75	2.68	1.32	1.31
	Stdev	0.40	0.57	0.79	0.06	0.03	0.13	0.13
D	TW305	771.43	60.91	−50.89	3.28	3.24	1.57	1.57
D	TW305	776.17	91.20	−80.11	3.35	3.30	1.61	1.63
D	TW305	771.57	69.06	−57.45	3.26	3.18	1.60	1.59
	Mean	773.06	73.72	−62.81	3.30	3.24	1.59	1.60
	Stdev	2.70	15.67	15.33	0.05	0.06	0.02	0.03

Table 6
LLS summary for 300 mm epi wafers

Lot average	Composite DFW 0.09–0.20, DFN 0.2–1.0			
date of delivery bin size (μm)	3/98 B(TW305)	12/97 C(TW305)	12/97 D(TW305)	6/98 C(TW305)
0.09–0.10	817	43	80	21
0.10–0.12	76	56	62	15
0.12–0.14	12	18	28	5
0.14–0.16	13	16	27	3
0.16–0.18	9	13	22	2
0.18–0.20	4	13	16	1
0.20–0.30	2992	48	122	9
0.30–0.50	92	45	100	6
0.50–1.00	53	57	128	3
total > 0.09	4067	308	585	64
total > 0.16	3149	174	388	21
> 0.09/ > 0.16	1.3	1.8	1.6	3.5
0.09–0.12	894	99	142	35
200 mm equivalent				
0.09–0.12	397	44	63	16
DFW haze (ppb)	116	74	213	98

for epi wafers would prevent detection of LLS smaller than 0.085 μm with the current P1 capability due to the saturation of PMT.

4.4. Epi defect inspection by scanning electron microscopy (SEM)

Physical position information of LLS detected by surface scanner P1 can be readily exported as ".tff" files to several commercially available SEM inspection stations for further investigation. Due to limitation of current positioning accuracy, only LLS larger than 1 μm were examined and shown in Figs. 8 and 9. Fig. 8 shows two typical structural defects found in epi wafers from supplier C and D. The crater-like defect shown in Fig. 8(a) has a depth consistent with the epi thickness measured by FT–IR, indicative of retardation of epitaxial growth due to probably non-optimized cleaning procedure. Fig. 8(b) shows a stacking fault-type growth defect with the length of each side ~4 μm. Fig. 9(a) presents an embedded epi defects from supplier B. For this particular defect, considerable size discrepancy observed between dark-filed narrow channel and dark-field wide channel. In narrow channel this defect was counted as an area defect with the size more than 20 μm, while in wide channel it was classified as a LLS in sizes 1 to 5 μm. The shape of the defect, which determines how the light was scattered into detectors, accounts for the observed size discrepancy. Fig. 9(b) shows a defect standing on top of the epi surface with part of its bottom melt into the epi layer. This defect appears to be a typical polymer-type deposition occurred at the end of epitaxy growth.

From the relatively high density of defects (> 20/wafer for sizes > 1 μm) and the types of defects identified we believe that the majority of 300 mm epi wafers suppliers are still in development phase and the quality of current 300 mm epi wafers will still need to catch up with the standards set by state-of-the-art 200 mm counterpart.

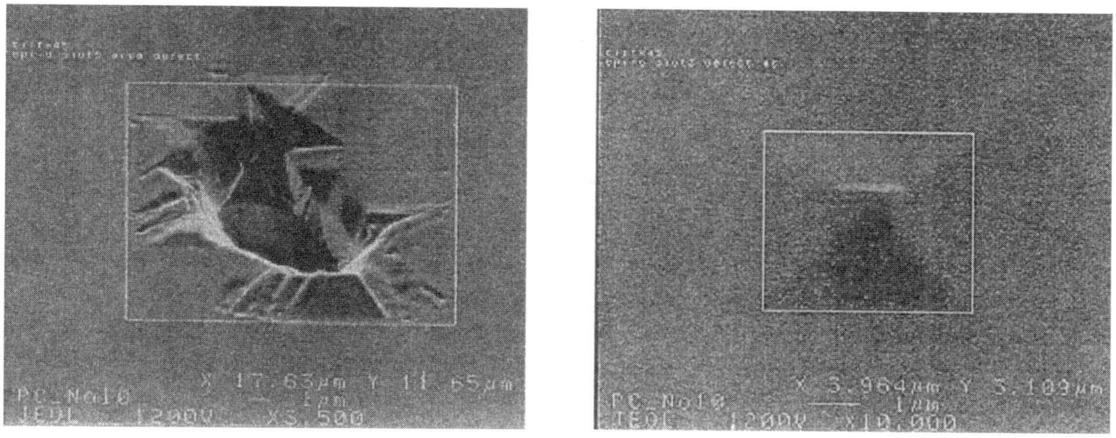

(a) (b)

Fig. 8. Structural defects observed in wafers of suppliers C, D.

5. Conclusion

Promising silicon characteristics comparable to state-of-the-art 200 mm wafers was demonstrated on 300 mm polished wafers. LLS with sizes down to 0.09 μm was shown much better controlled in 1998 compared to data collected last year. The early 300 mm epi wafers have exhibited admirable epi thickness control down to 0.15 μm, superior suppression of COPs in the range of 0.09–0.16 μm and cometitive site flatness performance. However, current 300 mm epi wafers have a worse larger particle readings than polished wafers primarily due to epitaxial growth-related issues. More efforts on controlling structural defects introduced during epi growth are required to improve the overall surface defect density.

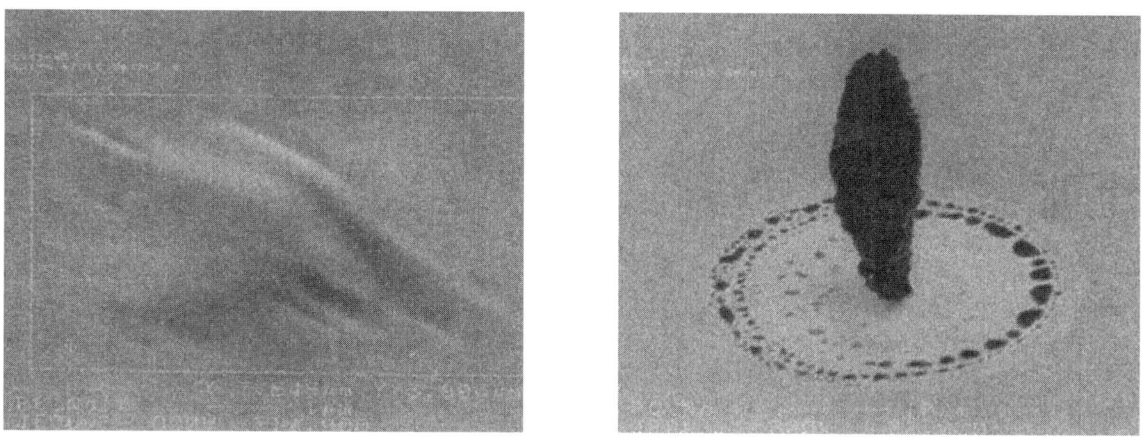

(a) (b)

Fig. 9. Embedded and polymer-like defects found in epi wafers.

Acknowledgements

We appreciate the assistance of Carolyn F.H. Gondran and Chris Sparks in facilitating the 300 mm wafer analytical measurements, Jeannette Zeagler for general and SEM support. Special thanks also go to major silicon suppliers who generously provide their wafers for the characterization. We also appreciate the great help from On-line Technology for measuring the epi thickness and ADE for the work of flatness measurement.

References

[1] F. Robertson, A. Allan, 300 mm Conversion Timing, Future Fab 1 (3) (1997) 27–33.
[2] H.R. Huff, D.W. McCormack, C. Au, T. Messina, K. Chan, R.K. Goodall, Current Status of 200 mm and 300 mm Silicon Wafers, in: ISSDM 1997 Japan, 1997, 456-457, 575 Conference Abstracts.
[3] SIA 1997 Roadmap, Semiconductor Industry Association, 181 Metro Drive, Suite 450, San Jose, CA 95110.
[4] Peter Wagner of Wacker Siltronic Corp., private communication.
[5] C. Au, T. Messina, R.K. Goodall, H.R. Huff, Characterization of 300 mm Silicon Polished Wafers, in: Proceedings of the 8th International Symposium on Silicon Materials Science and Technology, 1998, p. 641.
[6] SEMI Int'l Stds. 1997-Materials M1-97, 20, 1997.
[7] R.K. Goodall, H.R. Huff, Wafer Flatness Modeling for Scanning Steppers, SPIE 2725 (1996) 76–84.
[8] SEMI Int'l Stds.-Materials M1-(rev. 98, to be published).
[9] SEMI Int'l Stds. 1997-Materials M1.15-97 (1997).

ELSEVIER Microelectronic Engineering 45 (1999) 183–190

MICROELECTRONIC ENGINEERING

Three hundred-mm wafers: a technological and an economical challenge

Holger Dietrich[a,*], Werner Bergholz[a], Steffi Dubbert[b]

[a]*Siemens AG, P.O. Box 80 1709, D-81739 Munich, Germany*

[b]*Siemens Microelectronics Center GmbH and Co. OHG, P.O. Box 100940, D-01076 Dresden, Germany*

Abstract

For the transition from 200- to 300-mm wafers, in addition to the technological challenges, there is the equivalent cost challenge to be met. The ultimate target is cost parity per unit area of silicon between 200 and 300 mm. The major blocking point to be cleared appears to be crystal pulling. The question of cost of crystal pulling is mainly related to the large crystal weights (> 300 kg) and to the issue of crystal defects (crystal originated pits (COPs); voids in the Si crystal) generated by crystal pulling. An alternative route to address the defect problem is to use epitaxy to create a defect-free layer of Si for the active device regions. Close co-operation between wafer manufacturers and users is absolutely necessary to address these key issues. Furthermore, to contain cost, wafer makers and device manufacturers have to work together intensively to implement standardisation wherever possible and to avoid cost driving overspecifications (e.g. for front and backside particles or flatness).

1. Introduction

In the silicon industry, past transitions to larger wafer diameters have been driven mainly by one device company. As a result,

- the high cost of this change has put heavy burdens on that particular company;
- the new wafer parameters for the larger diameter were set by the first company, but the followers did not necessarily choose the same. For 200-mm wafers, e.g. for some time two wafer thicknesses (735 and 725 μm) existed, causing unnecessary cost at the wafer manufacturers.

As for previous transitions, the major driving force for the transition from 200 to 300 mm is to reduce the cost per manufactured device. It has been estimated that the migration from 200- to 300-mm wafers has the potential to reduce overall cost per chip by 30% [1]. Therefore, there is

*Corresponding author.

0167-9317/99/$ – see front matter
PII: S0167-9317(99)00135-5

consensus that these expensive mistakes of the past made in the course of the transition to a larger wafer diameter are to be avoided this time. Furthermore, it is clear that the transition would be too expensive to be supported by one company only and that a standardisation of wafer parameters should be achieved from the very beginning.

Therefore, two consortia have been formed (I300I, at Sematech (Austin) American, European, Korean and Taiwanese device houses) and SELETE (Japanese companies) to spread the cost of pre-competitive evaluations and facilitate the definition of standards. Furthermore, there is also consensus that an important factor for keeping cost under control is to avoid cost driving overspecifications.

In the course of the first exploratory work it quickly became obvious that as the diameter increases the trend for a larger price per unit area Si would be even more pronounced for 300- relative to 200-mm wafers. This disadvantage, of course, must be overcompensated by other cost advantages to result in an overall reduction in cost to produce a device. For many companies, the first devices to be manufactured on 300-mm wafers will be DRAMs where the cost advantage is absolutely essential, otherwise there will be no incentive to implement the transition in mass production.

As in the previous transitions, the step to a larger wafer diameter involves many technological problems (equipment and process related) which have to be addressed. This, of course, applies for the manufacturing of both wafers and devices.

So, in addition to the technological challenge, there is an equally important cost challenge for the introduction of 300-mm wafers. Only if both are met, the transition will materialise.

It is the purpose of this paper to elucidate that the key to meet these challenges is an intensive co-operation between the wafer manufacturer on the one hand and the device manufacturer on the other hand. To highlight this point, the device manufacturing aspects which call for the interaction between wafer makers and users are printed in italics.

2. Crystal pulling: a 3-D challenge

The increase in wafer diameter by a factor of 1.5 has the following major implications:

- For an equal yield of useful wafers relative to the amount of poly Si put into the quartz crucible, all other things being equal, the weight of the crystal to a first approximation has to grow with the third power of the diameter, i.e. by a factor of approx. 3.4, since the height of the end cones grows proportional to the diameter. Compared to a typical *charge weight* of 90 kg for 200-mm wafers, this amounts to 300 kg!
- The *dissipation of heat* is proportional to the volume of the crystal. Since the main heat transport is by heat conduction (which slows down with the square of the distance covered), the pulling rate has to be slower than for 200-mm crystals.
- The *larger volume of the melt* is more difficult to control (more intensive convection, longer time constants). Furthermore, a considerably larger crucible is required.
- The formation of *crystal defects* and incorporation of impurity atoms (*dopants and oxygen*) into the larger crystal volume is significantly altered, due to the different boundary conditions for diffusion and crucible/crystal geometries.

From these *volume-related, i.e. 3-D physical phenomena*, the most difficult technological and most severe cost challenges result:

- It is not clear whether the *thin neck* technique to implement dislocation free crystal growth (diameter typically 3 mm) can really be extended to crystal weights of the order of 300 kg and possibly beyond. One of the seven large Si suppliers holds a patent for a support mechanism which engages into a secondary neck grown at the initial phases of crystal growth. It remains, however, to be seen whether this technique can be developed to give a sufficiently high yield. If the thin neck problem cannot be overcome, this is a blocking point for 300-mm wafers. – To support the wafer manufacturers in the solution of this problem, the device manufacturers have to be prepared to support frequent changes in the volume of the charge weight which for mature technology is classified as a major change since it can have substantial implications on yields and failures in the device process.
- The slower pulling rate means a lower throughput in terms of crystal length. However, this effect is mitigated by the fact that the throughput in terms of unit area Si at a given pulling speed is increased by a factor of 2.25 due to the increased area of the 300-mm wafers compared to 200-mm wafers. The combination of the slow pulling rate with the larger footprint of the equipment and the higher cost for the large 300-mm crystal pullers results in a substantial cost driving factor. The suppliers can, in principle, envisage several 'forbidden' practices to make better use of the crystal and the poly. Again, close co-operation between the Si user and the Si maker is required to overcome such traditional 'no no's', such as the use of end cones for remelt. Furthermore, the extension of usable portions of the ingot can both be achieved through a critical examination of the real requirements of the oxygen content and the dopant concentration on the side of the user.
- It appears that there are no fundamental problems to obtain crystals with sufficiently good homogeneities of dopant or oxygen incorporation, although the detailed economics will still have to be investigated. – The device manufacturers' contribution is again to examine whether the homogeneity requirements can be relaxed to reduce the cost.
- The defect scenario could be another blocking point, because there is a double problem:
  In today's device grade prime CZ wafers there is a density of 10^5 to 10^6 cm^{-3} of COP, i.e. voids of diameters of a few 10 to some 100 nm. These result in a limitation of the gate oxide quality and reliability of the devices as the device design rules are approaching 0.18 or 0.13 μm (the design rule for mass production on 300-mm wafers). They might also cause problems with photo lithography and other failure mechanisms, like isolation defects, transistor parameters, etc. – Therefore a *reduction or complete elimination of these defects* is necessary!
  As the wafer diameter increases, the task to control and reduce the density of the defects during and after crystal growth becomes more difficult!

The implications of this point for device processing are most important: is it possible to achieve these targets by crystal pulling at reasonable cost or will it be unavoidable to migrate to epitaxy for prime wafers? Since epi is a costly process, this will be an additional severe cost burden. (This argument is only valid under the assumption that it is not necessary to use p^+ material to control latch-up). Obviously, this calls for most intensive co-operation between supplier and user, possibly also in the field of 200-mm wafers where stable device processes can be used to test improved gate oxide/reduced COP materials manufactured by different processes.

The longer time required for the melt down of the crystal and the longer pulling time lead to a reduced dislocation-free yield (all other things being equal), since there is a higher probability for events that lead to the generation of dislocations. To make matters worse, if a dislocation occurs it spreads out about the length of the crystal diameter, roughly making a 300-mm portion of the crystal unusable. Furthermore, there is an increased requirement for the thickness of the bubble-free layer of the crucible which is consumed during the crystal growth. Such crucibles in the ultimate required size of 32–36 inches (0.8–0.92 m diameter) are currently not available, this poses both a cost and a technological challenge! Crucible erosion itself increases the probability of particle generation which leads to dislocation formation, i.e. even higher cost. – Any change in the pulling process which increases the throughput and reduces the crucible erosion should, therefore, be supported by the device manufacturer. This can have implications, e.g. for the defect type and distributions and for the oxygen content, i.e. gettering ability of wafers (see also below).

- The motion of the melt can be controlled by the application of a magnetic field. This implies another cost-driving factor. It also has important consequences for the intrinsic gettering capability of the wafers, since the oxygen content is lowered considerably. There are certainly some device processes for which this could be a problem. This means that schemes are necessary that enable controlled oxygen precipitation for low-oxygen wafers. The device manufacturers can help by evaluating magnetic Czochralski (MCZ) material and the development of schemes for controlled oxygen precipitation also for smaller wafer diameters to support and accelerate the development for 300 mm.

These and other factors of smaller importance make crystal pulling the prime cost driver for 300-mm wafers. It is not entirely clear at this point whether the technological challenges can be met at a cost which still results in a sufficiently lower total cost for producing devices on wafers on 300-mm wafers.

3. Wafering: a 2-D challenge

Contrary to crystal pulling, the main issues of scaling up the wafer processing are obviously *area-related* (i.e. 2-D).

Generally speaking, the task to scale up wire saws, lapping or grinding equipment, edge-rounding apparatus, furnaces, polishing and cleaning equipment is a comparatively simple engineering exercise and has been essentially accomplished. Very similar to processing of wafers in device processes, the quality area simply has to be extended to a larger diameter. The prime concerns are economical ones: throughput (trend to more single wafer processes as the diameter is increased), footprint, uptime etc.

There are two remarkable technical developments, the first of which is mainly cost driven (wire sawing and grinding), the second one (double-side polishing) has both technology and cost aspects:

- *grinding:* compared to lapping, the single-wafer grinding technology is more expensive than the batch process lapping, however, this is overcompensated by a reduction in damage depth, which means that less of the 'precious' single-crystal silicon is consumed in etching (compare also Section 1). As an additional bonus further cost is obviously saved by less consumption of chemicals. The resulting improvement in wafer geometry comes free since less etching removal means less degradation of the flatness by etching. The exclusive use of wire saws instead of inner diameter-saws is the consequence of a 200-mm development/experience, namely that wire sawing generates the same level of warp at a significantly reduced consumption of silicon by the cut and again a smaller damage depth. – A smaller etch removal should be carefully *optimised* observing potential effects in the device process, such as particle generation, dislocation formation through residual damage, increased wafer breakage etc. Obviously, a very good feedback loop between manufacturer and user is required for this kind of exercise.

- *double-side polishing:* this was pioneered for smaller diameters, in particular 200 mm. It was introduced because of its larger potential for flatness capability for comparable machine parameters (stiffness, accuracy of surfaces and alignment of axes etc.).
  In the course of investigating the standard set of wafer parameters it was further shown that an additional advantage of a double-side polished wafer in combination with a polished edge is the reduced generation of Si particles by contact with chucks and transport systems and the higher robustness of such wafers with respect to slip generation in thermal processes. – Clear-cut evidence for the advantages of a polished edge are again the result of a very intensive co-operation [2]. If such evaluations had not existed, the definition of the polished edge as a standard feature for the 300-mm wafer might not have happened right from the beginning.
  There is also a small cost advantage since the expensive waxing and dewaxing processes are omitted.
  The polished back surface implies that automatic particle inspection is possible. As it can be measured, there is a trend to put it into the specification which immediately generates the need for *edge handling equipment*. For some processes *this can be a severe cost driver.* – Hence it has to be checked very critically by tests in device processes whether the specification of the particles on the wafer back surface really generates enough added value to overcompensate the increased cost!

Since both double-side polishing and grinding are comparatively new technologies, we are only at the beginning of the learning curve which will at least initially contribute to cost, but which does not appear to be a particular technology or cost risk. – This development should be supported by the device manufacturer through a correlation between yield results and process improvements/cost reduction steps in the wafering process.

Although there are first trends to move to double-side polishing also for 200 mm due to the more and more stringent demands for improved wafers, it appears doubtful whether this will become mainstream for the smaller diameter: the use of double-side polished and single-side polished wafers in one device line is possible, *but will generate many practical problems which could affect the economics of a mass production device line quite substantially!* Furthermore, the introduction of substantial double-side polishing capacity at wafer manufacturers could result in high cost for investment and lower utilisation of existing single-side polishing capacity. – As for 300 mm, a critical examination also for 200 mm in co-operation between maker and user of wafers is required to fathom how critical the flatness requirements are in reality and whether a required improvement of the wafer parameters is really warranted by the benefit. In this respect, *the timing* is also of critical importance: if the tightened flatness requirements are gradual, continuous process improvement for the existing single-side polishers can keep up with the device process requirements.

4. Epitaxy: – a 1-D challenge

There does not seem to be a true technological challenge for 300-mm epitaxial wafers: there are at least two working models of 300-mm epi reactors from the two major manufacturers of epi reactors and there have been announcements of first prime-like wafers which have been produced.

The real challenge of epitaxy is its higher price compared to polished wafers [3]. It has been projected that in the long run the price for 300-mm epi wafer will exceed that of a polished wafer by up to $200. It is quite clear that such a large price adder is not acceptable for low price mass product devices.

Assuming that polished prime wafers can be produced at a reasonable cost (compare Sections 2 and 5) the major challenge for epitaxy is to reduce the cost by a significant amount. This can be achieved by:

- A higher throughput, i.e. a thinner epi. The final thickness for the demands of 0.18- or 0.13-μm devices are to be determined. Hence epitaxy for 300-mm prime wafers is to a large degree a *1-D challenge*.
- In addition, a number of other factors to reduce the cost of epi need to be addressed, i.e. the deposition rate, the epi yield (both by improvement of the defect density of the reactors and avoiding overspecification).
- It is necessary to optimise reactors for high throughput (handling systems, etc.) and low cost for media consumption, uptime, footprint, etc.

Such developments again have to be accompanied by device evaluations to find an optimum for total cost.

It is predicted that if COP-free prime wafers can be manufactured at a reasonable cost [4] then epitaxy for mass-production devices will be attractive if either there are significant yield advantages or the price adder for epitaxy is not of the order of 30–50%, but 10% or below for a prime wafer.

Whether epi or polished CZ wafers will be the main stream prime wafers is one of the most important decisions to be reached in the course of the ramp-up of 300-mm technology.

Obviously, to address this key problem, both the close co-operation of suppliers and users and an evaluation strategy involving also 200 mm is essential, since only on 200-mm wafers there are device processes available with a sufficiently stable yield. In this connection, the evaluation of low COP material and cost optimised epi wafers will generate the necessary data for the decision which route to follow for future 300-mm prime wafers.

5. Characterisation of wafers and device processes

As mentioned before, due to the reduced design rules, there is a demand for monitoring smaller particle diameters and not only to produce, but also to measure smaller and smaller values for the local flatness.

For the detection of particles there is at present the complication that in addition to particles laser scanners also detect COPs, the number of which is by far larger than the typical number of particles. There are, however, developments underway that allow a separation between particles and COPs up to a reliability of about 80–90%.

A further challenge is the extension to smaller particle diameters and a standardised definition of extended defects such as scratches etc. As for process capabilities versus tightness of specified parameters, a critical examination of need versus cost is necessary on the side of the user.

The COP issue is also very critical for potential cost for monitor wafers: for 200-mm wafers, quite complicated reusing (involving chemical film stripping and cleaning) and reclaiming (i.e. repolishing of wafers) schemes have been implemented to bring down the cost of process monitoring in device lines to control the exorbitant cost for monitor wafers. Obviously, more stringent requirements for critical particle diameters in equipment testing will bring down yields for internal recycling and reclaiming. – Also in this field, good co-operation between wafer supplier and device manufacturer can help to bring down cost of the 300-mm device line: wafers which have been optimised for low COPs in a way not compatible with prime wafer requirements are currently being developed.

The process capabilities and flatness requirements are approaching the 100-nm range, which is the limit of the technology of eddy current-based flatness testers. Furthermore, a smaller probe size than 2–4 mm is required to increase the spatial resolution and to obtain reliable measurements close to the wafer edge.

To address both problems, optical systems have been developed, which make use of the fact that 300-mm wafers are double-side polished, hence it is possible to measure the flatness by interference.

For particle, flatness and other characterisation tools it is essential for cost effectiveness to agree on standardised recipes; optimised parameters avoiding overspecification at as early a stage as possible. Obviously, good communication and co-operation again is the key to success.

6. Standardisation

Standardisation is one of the main issues of the two consortia for the conversion from 200 to 300 mm. The precompetitive co-operation to achieve a cost reduction by the use of standards reaches deep into the production process. With the conversion to a new wafer size there is a chance for a fundamental improvement in all fields of the IC production.

A strong potential for cost reduction that is being addressed is the standardisation of wafer parameters that are not ruled by products or processes at the IC makers such as the wafer thickness (see above), the shape of the edge profile (currently under discussion), laser marking, etc. The 200-mm world currently lacks many of these standards which leads to the big variety of different process steps at the wafer makers causing mostly unnecessary costs.

Another field where standardisation is expected to result in a significant cost reduction is the use of one standard reusable shipping box for the 300-mm wafers. Again for 200-mm wafers many wafer makers use their own shipping box and shipping concept. The reuse of shipping boxes and packing material is still in its infancy for 200-mm wafers.

7. Summary and outlook

One of the most important issues in the migration from 200- to 300-mm Si technology is the containment of wafer manufacturing cost, the final target being cost parity per unit area of silicon between the two diameters.

At present it is not clear how fast and whether at all the major blocking point crystal pulling can be cleared. We have pointed out extensively that the reduction of wafer manufacturing cost is critically dependent on an intensive co-operation between wafer suppliers and users, since technological and cost issues are connected in a complex way. Moreover, in many cases it is possible/necessary to use experience and technologies of 200-mm wafers.

A more general possibility for using synergies between 200 and 300 mm is the following 'copy exactly scheme' followed through by SC300, the joint venture between Siemens and Motorola. The first product manufactured on 300-mm wafers will be a 64M DRAM which is a volume product of the Dresden 200-mm line. Since the 300-mm pilot line is integrated into the 200-mm cleanroom, there is the potential to have an exact copy not only for the process and the wafer requirements, but also for the process media and cleanroom procedures. Thus, there is not only a risk and cost minimisation through this strategy, but it is also possible to have a direct comparison of the total cost for making devices on both wafer diameters.

- The key points to meet both the cost and technological challenges are:
- intensive co-operation between wafer maker and user;
- standardisation of edge profile, laser marking, shipping boxes, thickness, etc.;
- avoidance of overspecification, appropriate timing for tightening of spec parameters such as local light scatterers (COP/particles), backside particles, local flatness, range of the oxygen content, resistivity range, crystal type (vacancy–interstitial–ring), etc.;
- obtain more insights into the *epi vs. non-epi for prime wafers question* soon

The successful migration to 300 mm will to a large degree depend on whether these challenges can be met in time or not.

Acknowledgements

The authors are grateful to Dr Hermann Jacobs for intensive discussions.

References

[1] R. Goodall, in: H. Huff, U. Gösele, H. Tsuya (Eds.), Semiconductor Silicon 1998, Electrochem. Society, Pennington, NJ, 1998, p. 1303.
[2] R. Winkler, W. Hanke, in: H. Huff, U. Gösele, H. Tsuya (Eds.), Semiconductor Silicon 1998, Electrochem. Society, Pennington, NJ, 1998, p. 1325.
[3] A. v Zitzewitz, Markt Tech. 5 (28) (1998).
[4] MEMC Press release on APEX material, April 13, 1998.

ELSEVIER

Microelectronic Engineering 45 (1999) 191–196

MICROELECTRONIC ENGINEERING

Discrimination of particles and defects on silicon wafers

F. Passek*, R. Schmolke, H. Piontek, A. Luger, P. Wagner

Wacker Siltronic AG, P.O. Box 1140, D-84479 Burghausen, Germany

Abstract

Angle-resolved light scattering measurements of particles and crystal originated particles (COP) point out ways for discriminating different types of defects on Silicon wafers. Particles and COP display a significantly different light-scattering behavior. Comparing the intensity of light scattered into the different solid angles of the two detection channels of the KLA-Tencor SP1 is sufficient for distinguishing particles from other defects. The share of particles identified correctly by SP1 data analysis depends on the defect size and on the wafer material and is >85%.

1. Introduction

Discriminating different kinds of defects on silicon wafers is required by the 1997 edition of the National Technology Roadmap for Semiconductors (NTRS, [1]) which distinguishes localized light scatterers (LLS) and particles on polished Silicon wafers.

Standard defect discrimination and classification is performed by optical, electron, scanning tunneling and atomic force microscopes or other analytic tools. Such measurements are time consuming and allow only spot checking. Therefore these defect classification methods do not meet the requirements for final inspection in manufacturing of polished Si wafers. Final inspection is commonly performed by using scanning surface inspection systems (SSIS). The entire wafer surface is scanned by a well focused laser spot and light scattered by surface defects and roughness is collected and detected by an appropriate optical and detector system in a dark field mode. Scattered light intensity locally significantly larger as compared to the one due to surface roughness is identified as localized light scatterer (LLS) [2]. Recent theoretical studies demonstrate that the angular distribution of the light scattered by defects on Si wafers contains information about the shape of the scatterers [3]. Simulations and experiments of light scattering on arbitrary shaped pits [4] and particles [5,6] reveal significant differences in the angular pattern of scattered light. Therefore, a discrimination regarding the specific nature of a defect appears to be basically possible. However, most of the presently installed SSIS are equipped with only one dark field channel and can not record angular

*Corresponding author. Tel.: +49-8677-83-6560; fax: +49-8677-83-2640.
E-mail address: friedrich.passek@wacker.de (F. Passek)

0167-9317/99/$ – see front matter
PII: S0167-9317(99)00145-8

dependency. Recently SSIS with two or three dark field channels for light scattered in different solid angles became available and it needs to be checked whether data of two or three dark field channels are sufficient for defect discrimination.

The present paper reports the results of investigations on light scattering of particles and pits as recorded by an angle-resolved light-scattering system (ARLS) [7] and on the capability of a SSIS, KLA-Tencor SP1 [8], equipped with two dark field channels at fixed positions, to discriminate particles from other defects on final polished Si wafers.

2. Experimental

2.1. Angle-resolved light-scattering system

The ARLS is a non-commercial system equipped with a Argon ion laser with a wavelength of 488 nm, a movable detector system and a high precision sample stage. The diameter of the normally incidenting laser beam on the sample is 40 μm, the aperture angle of the detector is about 2°. The angle dependence of scattered light was investigated by recording the scattering data as function of the detector position which can be varied between 27 and 75° with respect to the incidenting beam. Phase-sensitive amplification and detection was used to improve the signal-to-noise ratio. Light scattered from a defect was separated from that caused by roughness by subtracting the angle resolved intensity obtained from an adjacent, defect-free reference area from the data recorded in a measurement of a defect.

2.2. Surface scanning inspection systems

A SSIS KLA-Tencor SP1 was used in the present investigation. It is also equipped with an Argon ion laser with a power of 20–30 mW. The laser beam again is normally incidenting on the wafer surface. The spot size on the wafer is 20×50 μm in the low throughput mode used in the present investigation.

The scattered light is collected by a lens and an ellipsoidal mirror system, respectively, and focused in two separate photon multiplier tubes at fixed positions. These two dark field channels, the so-called dark field narrow channel (DFN) and dark field wide channel (DFW) integrate light scattered in narrow angles, i.e. about 5–20° with respect to the incoming beam, and in wide angles between about 30 and 70°, respectively. Therefore, a really anglular resolved measurement is not performed by the SP1, but only approximated by using two fixed angles and integrating over substantial solid angles.

The amount of scattered light collected by different SSIS or different channels depends on the arrangement (solid angle, scattering angle) of their optical systems. Therefore, comparing data of different SSIS requires to eliminate the characteristics of the detection systems. The international accepted standard method to perform this elimination is based on calibrating SSIS with polystyrene latex spheres (PSL) [9]. This allows to relate light intensity with a latex sphere equivalent size (LSE) equivalent to a diameter of PSL spheres and the DFN and DFW of the SP1 were calibrated accordingly. The detection ranges, expressed in terms of LSE sizes, were set up between 90–300 and 110–300 nmLSE in the present investigation for the DFW and DFN, respectively.

2.3. Material

Eight- as well as 12-inch wafers with an electrical resistivity of two different ranges were investigated. Material A was doped with 2.8×10^{15} to 1.5×10^{15} atoms/cm^3 Boron while material B had a boron concentration of 2.0×10^{19} to 8.5×10^{18} atoms/cm^3. Wafers of material B have been shown to contain less crystal originated particles (COP) as compared to material A [10].

2.4. Defect identification

Correlating the scattering behavior of LLS with their specific nature requires an unique identification of the scattering objects. A large number of well characterized defects has to be investigated to obtain statistically significant results. Therefore, the standard defect characterization method with time-consuming optical or scanning microscopes is not appropriate for performing this job. Cleaning the wafers between two successive measurements is an alternative approach. It allows to separate the LLS detected in stationary (S)-LLS and non-stationary (N)-LLS. S-LLS are the LLS detected at identical coordinates during both measurements. N-LLS are, correspondingly, LLS that have not been observed in the second measurement after cleaning.

S-LLS can be assumed to be crystal-related surface defects (COP, [11]) or damage provided the cleaning process removes particles effectively. N-LLS can be particles removed in the wafer cleaning or stationary defects not detected in the second run due to the finite detection probability of the SP1. The probability for a S-LLS not to be detected in the second measurement depends on the LLS size and drops from 80% at the detection threshold of 90 nmLSE to 7% at 95 nmLSE and to 5% for larger diameters.

3. Results and discussion

3.1. ARLS

The angle-resolved scattered light intensity of particles and COPs were normalized to similar intensities at small angles for comparing the angular-dependent variations. This normalization was performed using polynoms fitted to the ARLS data (Fig. 1). Sizes of 152 and 151 nmLSE were assigned by the DFN of the SP1 to the COP and particle, respectively, of which the ARLS data are shown in Fig. 1. The intensity curves show similar behavior for COPs and particles but differ significantly for angles $>35°$ indicating different scattering behavior of particles and COPs. COPs scatter less light in larger angles than particles. The difference between the intensity scattered by particles and by COPs increase with increasing scattering angle and has a maximum of 20% at about 60°. These experimental data demonstrate the potential of angle-resolved detection of scattered light, to discriminate particles from other defects.

3.2. SSIS

S-LLS were reported by the DFW to have significantly smaller sizes as compared to the DFN for 300-mm wafers of material B (Fig. 2). The size range for S-LLS displayed in Fig. 2 is 110–300

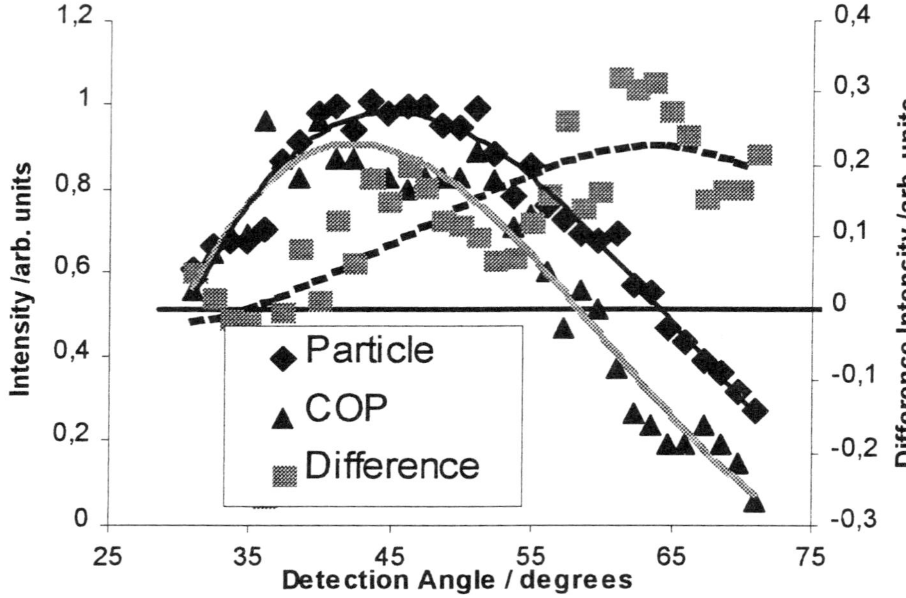

Fig. 1. Angle-resolved light-scattering data of a particle and a COP to which a size of 151 and 152 nmLSE, respectively, were assigned by the DFN of the SP1. The data were normalized to similar intensities at small scattering angles. The solid lines represent polynoms fitted to the data. In addition the difference between the intensity of the particle and the COP is shown.

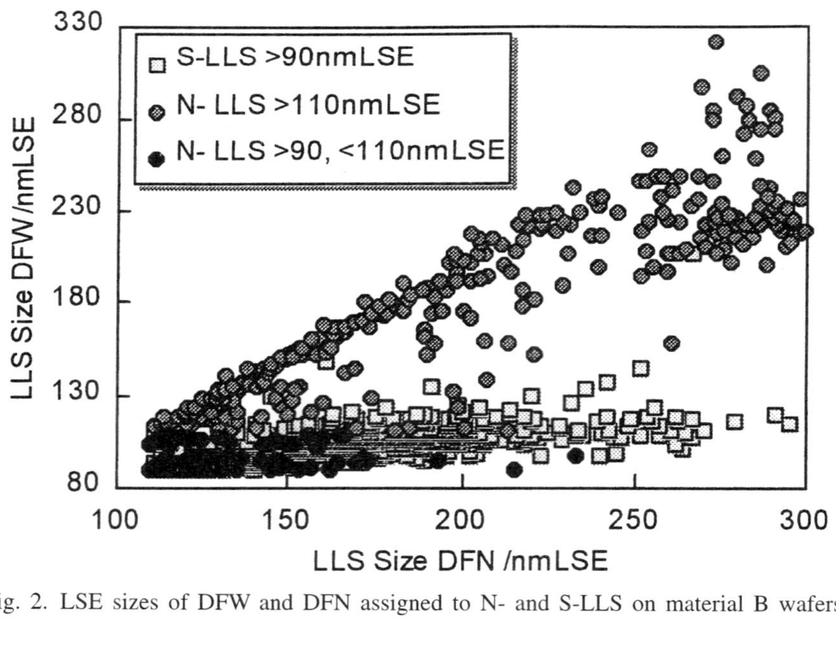

Fig. 2. LSE sizes of DFW and DFN assigned to N- and S-LLS on material B wafers.

nmLSE and only 90–120 nmLSE for the DFN and DFW, respectively. S-LLS above 120 nmLSE (DFW) appear not to occur on these wafers. The N-LLS can be divided in two groups corresponding to their different sizing behavior. N-LLS with a size >110 nmLSE DFW were reported by both detection channels to have nearly identical diameters while N-LLS < 110 nmLSE (DFW) were measured with significantly larger sizes by the DFN. N-LLS < 110 nmLSE (DFW) have a size close to the detection threshold of the SP1. The probability for an LLS with a size at the detection threshold to be detected in a first as well as in a second measurement is 20%. Therefore, it can be assumed that the majority of N-LLS < 110 nmLSE (DFW) are crystal defects (COP) and damage falsely identified as N-LLS.

The ratio R of the diameters assigned to an LLS by DFN and by DFW appears to be a better criterion for COP/particle discrimination than the size difference (Fig. 3). An average R of about 1.05 is obtained for material A as well as for material B, while R is significantly larger for S-LLS. However, the ratio between the DFN and DFW sizing for S-LLS depends on the wafer material. The size ratio is about 1.25 and 1.4 for S-LLS on materials A and B, respectively. The curves in Fig. 3 reveal the limits of the N- vs. S-LLS separation, too. The curves for particles – N-LLS- and S-LLS overlap. About 85% of the N-LLS, 90 and 95%, respectively, of the S-LLS were assigned correctly using $R = 1.15$ and $R = 1.25$ as separation threshold.

These data of S-LLS and N-LLS >110 nmLSE (DFW) confirm the ARLS results and demonstrate that a quite unique separation of particles is possible even by an angle-resolved measurement using only two fixed angles and solid angles, respectively.

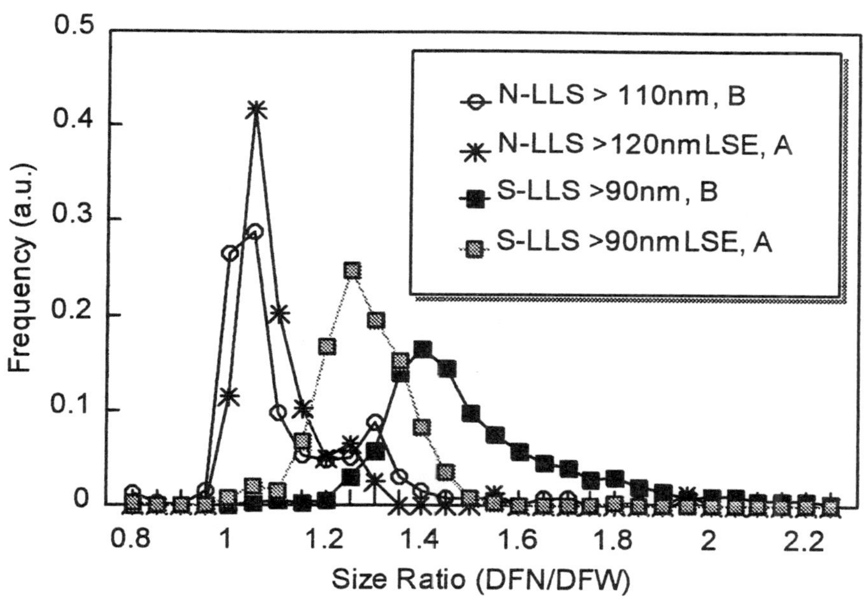

Fig. 3. Distribution of the ratio of the LSE sizes assigned to N- and S-LLS by the DFN and DFW of the SP1 for materials A and B wafers.

4. Summary

The light scattering behavior of particles differs from the one of crystal-related surface defects. This allows to distinguish particles from other defects not only with highly angle-resolved light scattering technique, but also by detecting light scattered in only two different solid angles. The probability for identifying particles correctly with the commercial SSIS, KLA-Tencor SP1, is between 85 and 90% and it depends slightly on the wafer material. The ratio of the diameters assigned to a defect by the two detectors is similar for large and small LLS and, therefore, this ratio is considered to be an efficient criterion for discriminating defects.

Acknowledgements

The support and assistance of H. Gerber, G. Vallouch, and C. Schnepf is gratefully acknowledged. This work was supported by the Federal Department of Education and Research of Germany under contract number 01 M 2973 A. The authors alone are responsible for the contents.

References

[1] National Technology Roadmap for Semiconductors (NTRS) of Semiconductor Industry Association (SIA) (1997).
[2] Semiconductor Equipment and Material International, SEMI M1-95 in Book of SEMI Standards, SEMI, Technical Publications (1995).
[3] Yu.A. Eremin, J.C. Stover, N.V. Orlov, Optic. Eng. (submitted).
[4] V.I. Ivakhnenko, C. Scheer, J.C. Stover, Proc. SPIE (1998) (submitted).
[5] V.I. Ivakhnenko, J.C. Stover, Yu.A. Eremin, Proc. SPIE (1998) (submitted).
[6] J.C. Stover, V.I. Ivakhnenko, C.A. Scheer, Proc. SPIE (1998) (submitted).
[7] R. Schmolke, D. Gräf, M. Suhren, R. Kirchner, H. Piontek, P. Wagner, Mater. Res. Soc. Symp. Proc. 442 (1997) 137.
[8] H. Altendorfer, G. Kren, C.T. Larson, S.E. Stokowski, Solid State Technol. (1996).
[9] R.S. Howland, C.T. Larson, Laser Focus World, 99 (July 1995).
[10] M. Suhren, D. Gräf, U. Lambert, P. Wagner, Electrochem. Soc. Proc. PV96–13 (1996) 132.
[11] J. Ryuta, E. Morita, T. Tanaka, Y. Shimanuki, Jpn. J. Appl. Phys. 29 (1990) L1947.

ELSEVIER

Microelectronic Engineering 45 (1999) 197–208

MICROELECTRONIC ENGINEERING

Alkaline cleaning of silicon wafers: additives for the prevention of metal contamination

A.R. Martin[a,*], M. Baeyens[b], W. Hub[c], P.W. Mertens[b], B.O. Kolbesen[a]

[a]*Institute of Inorganic Chemistry/Analytical Chemistry, Marie Curie Straße 11, D-60439 Frankfurt am Main, Germany*
[b]*IMEC, Kapeldreef 75, B-3001 Leuven, Belgium*
[c]*Siemens Semiconductors, Otto-Hahn-Ring 6, D-81739 Munich, Germany*

Abstract

The paper presents a new strategy for cleaning of silicon wafers. A novel class of chelating agents added to alkaline cleaning mixtures provides promising performance without negative effects such as metal redeposition due to residual metal contamination of the cleaning solution. The superior capability of the new cleaning process is confirmed by the results obtained from wafer surface metal analysis as well as from minority carrier lifetime and diffusion length measurements and gate oxide integrity tests. Particle densities and surface roughness are not influenced by the presence of the chelating agents in the cleaning solution. TOF–SIMS measurements do not indicate any deposition of chelating agent on the wafer surface. With this type of modified SC-1 cleaning procedure the acid SC-2 step used in conventional RCA cleans to remove the metals deposited in the preceding SC-1 step is unnecessary resulting in substantial cost savings with respect to chemicals, waste, equipment and space.

Keywords: Cleaning procedures; Chelating agents; Heterocyclic compounds

1. Introduction

In todays' mass production of advanced microelectronic devices wet cleaning processes still take up about 30% of all process steps. Therefore wet cleaning processes are an important field for cost reduction efforts. At present various cleaning solutions are in use dependent on the type of contamination which has to be removed, e.g. particles, metal ions, precious metals and organic matter. Particles constitute the most critical contaminants with respect to device failures and, hence, yield. Metal contamination can also give rise to substantial yield losses and device reliability problems due to, e.g., degradation of gate oxide integrity (GOI), increased leakage currents and retention time failures in dynamic random access memories (DRAMs) [1,2].

All cleaning solutions have more or less undesired side effects (Table 1). Since no single cleaning

*Corresponding author.
E-mail address: martin@chemie.uni.frankfurt.de (A.R. Martin)

0167-9317/99/$ – see front matter
PII: S0167-9317(99)00150-1

Table 1
Common cleaning solutions in semiconductor manufacturing

Term	Chemicals	Removes	Mechanism	Side effects
SPM	H_2SO_4/H_2O_2	organic residues (e.g. photo resist)	oxidation	particles, sulfate deposition
HF-dip	HF/H_2O	oxides, metals in silicon dioxide	dissolution of silicon dioxide	particles, outplating of precious metals
SC-1	$NH_4OH/H_2O_2/H_2O$	particles, organics	etching of silicon, oxidation	metal ions redeposition, e.g. Al^{3+}, Fe^{3+}, Zn^{2+}, surface roughness,
SC-2	$HCl/H_2O_2/H_2O$	metal ions	complexation, hydrolysis	particles (corrosion)

solution is capable to remove all contaminants simultaneously, a sequence of different cleaning processes is used in practice.

The most common cleaning sequence is the RCA clean [3]. It consists of two steps called SC-1 and SC-2, respectively. The first is a mixture of ammonia, hydrogen peroxide and water and is very effective in removing particles and organic matter, but causes metal deposition on the wafer surface depending on the metal content of the chemicals. Removal of this metal contamination is the aim of the SC-2 step, a mixture of hydrochloric acid, hydrogen peroxide and water. SC-1 might cause surface roughness due to oxygen bubbles from hydrogen peroxide decomposition sticking to the wafer surface. SC-1 has been optimized with respect to surface roughness and particle cleaning efficiency [4].

To meet the requirements concerning metal surface concentration in SC-1 cleaning the purity of chemicals has been increased continuously. However, those chemicals are more expensive and another drawback is that they are very sensitive to unintentional and cross contamination.

Optimization or replacement of the conventional RCA clean by more economic alternatives is of crucial importance for present and future semiconductor manufacturing with respect to cost reduction. Cost reduction can be achieved by either reduction of chemical consumption, waste, equipment and clean room area, provided that the cleaning efficiency for particles and light organics is maintained. Three different approaches are being followed in the literature: (1) diluted chemistry and reduced temperatures in RCA cleaning, e.g. diluted SC-1 (1/4/20) at 50°C; (2) innovative diluted acid cleaning solutions using diluted HF ('dHF') and ozone ('IMEC clean') [5] and dHCl in addition ('diluted dynamic clean' DDC) [6] and (3) optimizing the SC-1 solution by addition of chelating agents to prevent metal precipitation onto wafer surfaces and thus being able to omit the SC-2-step [7,8].

We pursued the third way to achieve lower metal surface concentrations using present sub ppb metal grade chemicals or to maintain comparable metal surface concentrations when using lower quality grade chemicals for existing technologies.

2. Chelating agents (CAs)

Manifold applications exist for chelating agents in science and technology and plenty of literature is available about complex chemistry [9].

Basic requirements for chelating agents used in silicon wafer cleaning comprise:

1. sufficiently high complex stability constants for the relevant harmful metal contaminants
2. high chemical stability of both the free chelating agent and its metal chelates in the strongly oxidizing SC-1 solution.

In Fig. 1 the pH-dependence of stability constants for EDTA (ethylenediamine tetraacetic acid) complexes of various metal ions is outlined [10]. In the pH-regime of 9.5 to11 where SC-1 cleaning mixtures are operating relevant harmful metals like Fe and Cu exhibit reasonably high stability constants. However, as will be shown in the experimental section the chemical stability of EDTA in SC-1 is not sufficient. Beside prevention of metal contamination, chelating agents further are used to stabilize technical grade hydrogen peroxide. Decomposition of hydrogen peroxide in SC-1 solutions gives rise to surface roughness and formation of light point defects (LPDs) due to oxygen bubbles sticking to the surface preventing silicon etching [4]. Therefore reduction of the decomposition of hydrogen peroxide in SC-1 would reduce surface roughness and the number of LPDs.

There are a lot of inorganic and organic compounds forming charge transfer complexes like ammonia, cyanide, acetic acid, citric acid, chatechole, choline and many more. Connecting those functional groups to basic organic backbones leads to multidentate ligands called chelates. The strength of the bond between a metal ion and a single functional group in a multidentate ligand is

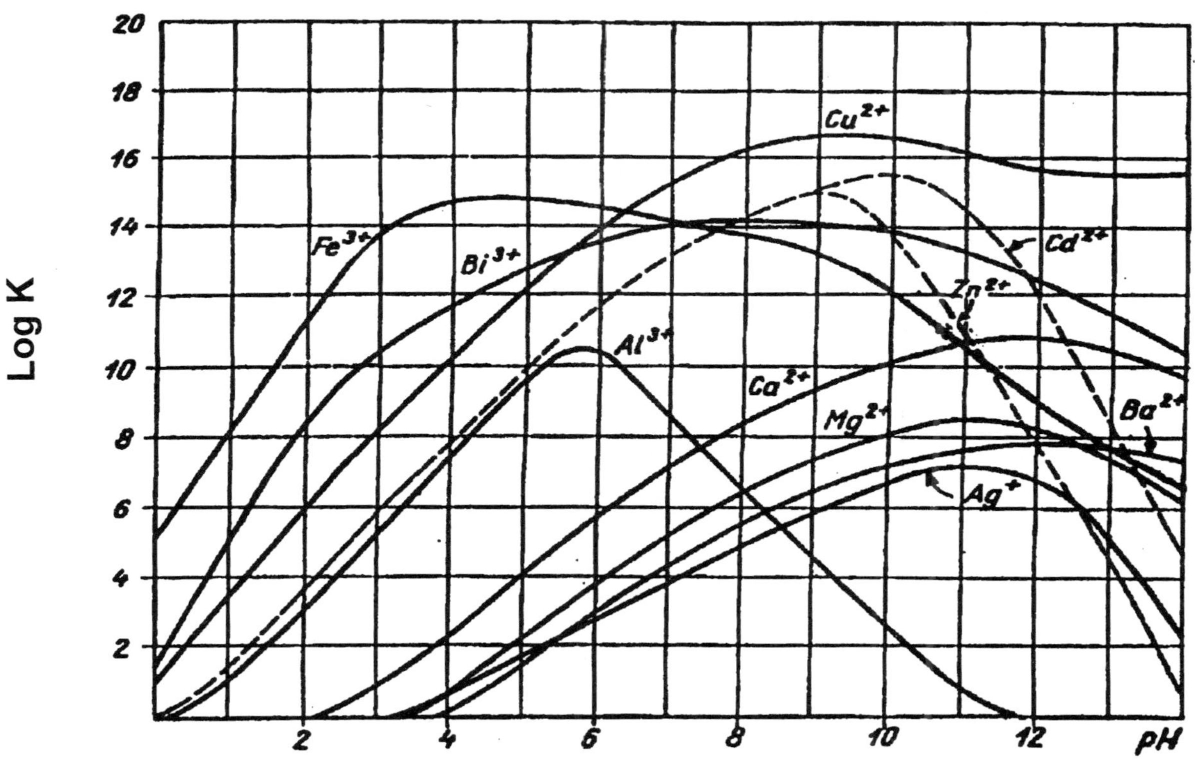

Fig. 1. Calculated conventional stability constants for EDTA complexes of various metal ions as a function of pH [10].

about the same order of magnitude as between the bond of this metal ion and the free ligand. The much higher formation constant of multidentate chelates is mainly attributed to the entropic effect ('chelate effect'). The chelate effect increases with the number of complex forming groups available in a molecule (Fig. 2).

Combination of different functional groups with various backbones offers a way to modify the properties concerning chemical stability, selectivity for different metal ions and complex forming constants. As backbones serve aromatics, saturated open chained or cyclic compounds. The latter two can contain heteroatoms like nitrogen or sulfur which form charge transfer complexes itself. Fig. 3 shows examples of chelating agents. The new class of chelating agents described in the present paper are heterocycles containing nitrogen with methylene phosphonic acid groups (cTRAMP) or acetic acid

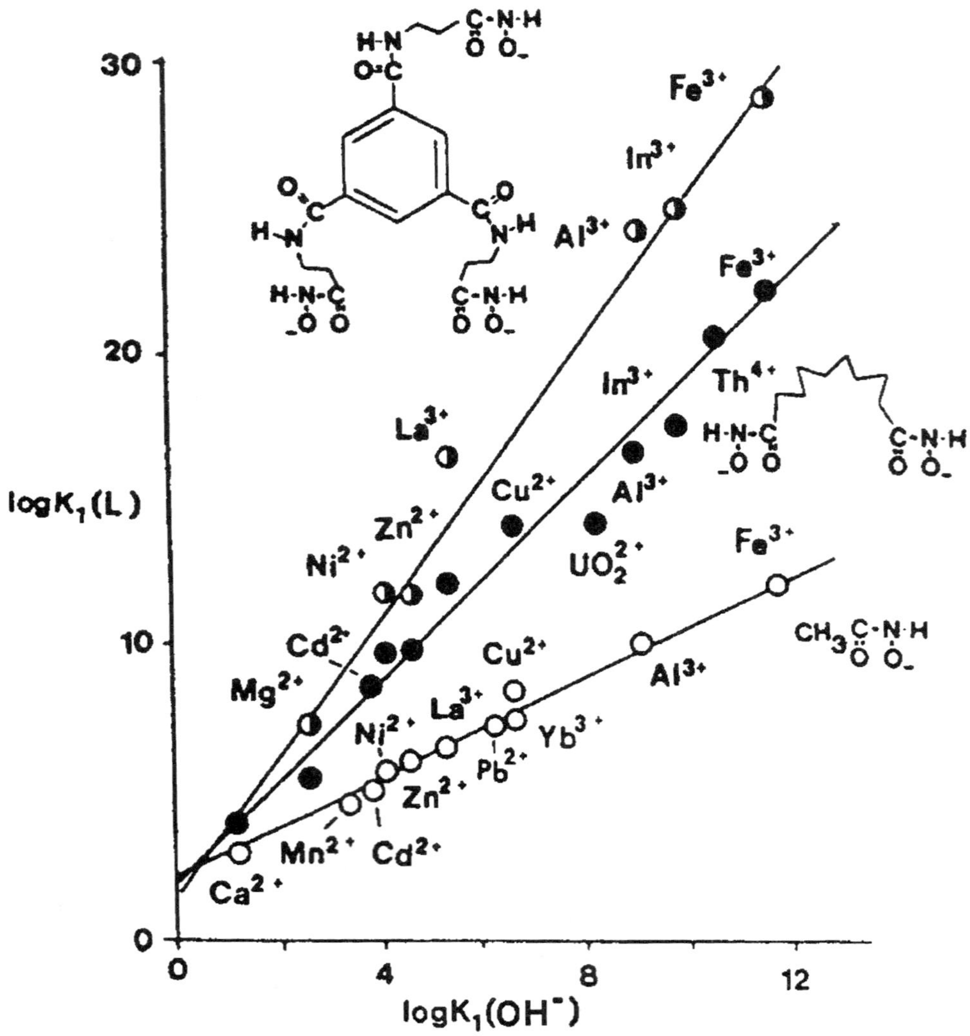

Fig. 2. Relationship between the formation constants of ligands containing negatively charged oxygen donor hydroxamate groups and log K_1 (OH^-) for the metal ions [9].

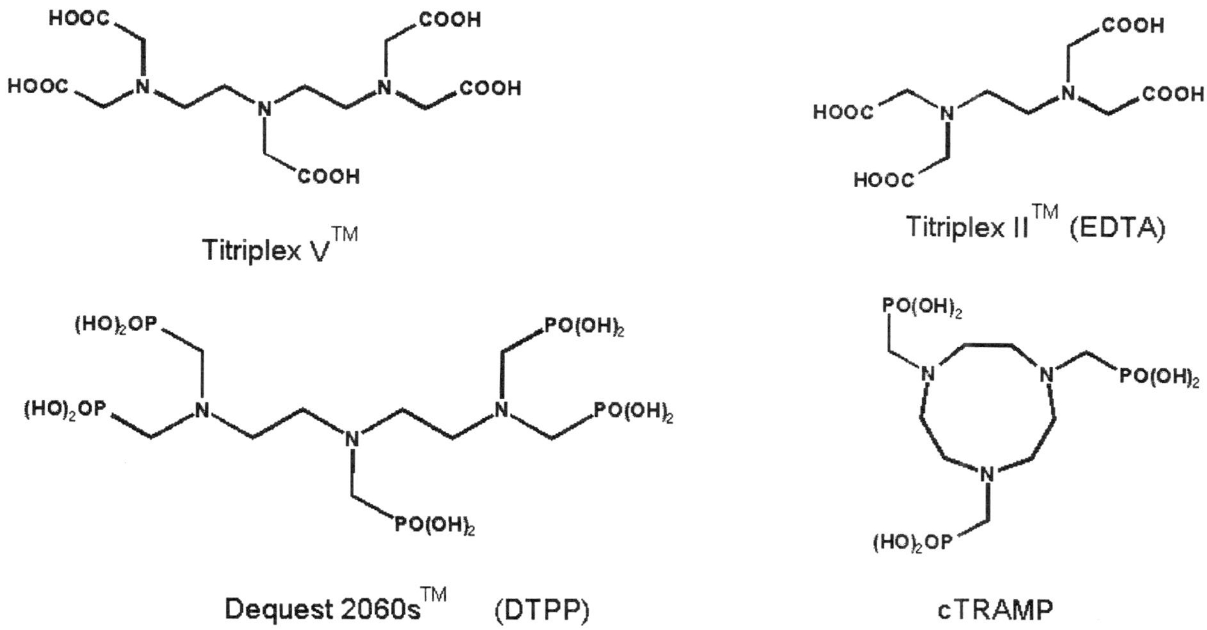

Fig. 3. Structural formula of chelating agents under investigation.

groups, respectively. They are compared with their analogues, well known open chained compounds with methylene phosphonic acids (Dequest 2060s™) and carboxylic acids (Titriplex V™, EDTA), respectively. Dequest 2060s is commercially available from Monsanto and is used to stabilize hydrogen peroxide solutions for bleaching and disinfection. EDTA and Titriplex are chelating agents used in quantitative analysis (chelatometry).

3. Experimental

For precipitation studies Ashland's Gigabit™ grade (metals <0.1 w-ppb) hydrogen peroxide and ammonia solutions were used ('GB'). Mixing ratios of 1/4/20 and 1/3/80 ($NH_4OH/H_2O_2/H_2O$) at 70°C and 50°C, with and without intentional metal contamination (1 w-ppb of each metal Ca, Fe, Ni, Cu, Zn) and with and without addition of 1 ppm of Dequest and cTRAMP, respectively, were employed. Standard boron doped EPI and CZ p-type wafers with (100) orientation and 150 mm diameter were immersed into the cleaning solution for 10 min. Metal surface concentrations of Ca, Fe, Ni, Cu, Zn were determined via Vapor-Phase-Decomposition Droplet-Surface-Etching Total-Reflection-X-Ray-Fluorescence Spectrometry (VDP–DSE–TXRF). For surface photo voltage (SPV) and ELYMAT measurements the metals deposited on the surface were driven into the bulk of the wafer by RTA (rapid thermal annealing) at 900°C. Via SPV iron bulk concentrations were calculated from the differences in photo voltages before and immediately after thermal annealing of the wafers at 200°C. Minority carrier life-time and diffusion length were determined by measurement of the photocurrent in an ELYMAT II.

The capability of preventing or slowing down hydrogen peroxide decomposition was proven by measuring the partial pressure of oxygen versus bath life time [11]. Wafers were immersed in pure SC-1 solution, in SC-1 spiked with 1 w-ppb of metals and in SC-1 with 1 ppm of CA (cTRAMP, diammonium–EDTA, Dequest) with and without metal spiking.

To ensure that the addition of chelating agents does not produce additional particles on the wafer surface, LPD measurements by TENCOR and CENSOR equipment were performed. TOF–SIMS measurements of wafers treated in SC-1 with Dequest 2060s and cTRAMP, respectively, were applied to check that the chelating agents do not adsorb on the surface of the wafer.

Gate oxide integrity (GOI) was tested by ramped voltage stress EBD and constant current stress QBD on wafers of cleaning experiments with and without metal spiking and with and without CA addition in comparison to several standard cleans: NCC, SC-1 (1/1/10), SC-1 (1/4/20) with and without addition of 1 ppm cTRAMP.

4. Results

4.1. Metal surface analysis

VPD–DSE–TXRF results of SC-1 cleaning without intentional metal contamination are given in Fig. 4. The detection limits of the method estimated according to DIN 32 645 are indicated also.

Dequest and cTRAMP are able to reduce iron surface concentrations below the detection limit of the method. Copper and zinc surface concentrations are only decreased by cTRAMP. EDTA is working in ten times higher concentrations only. VPD–DSE–TXRF results of SC-1 cleaning with intentional metal contamination are plotted in Fig. 5.

The new chelating agent cTRAMP prevents precipitation of Fe, Cu and Zn onto the wafer, while

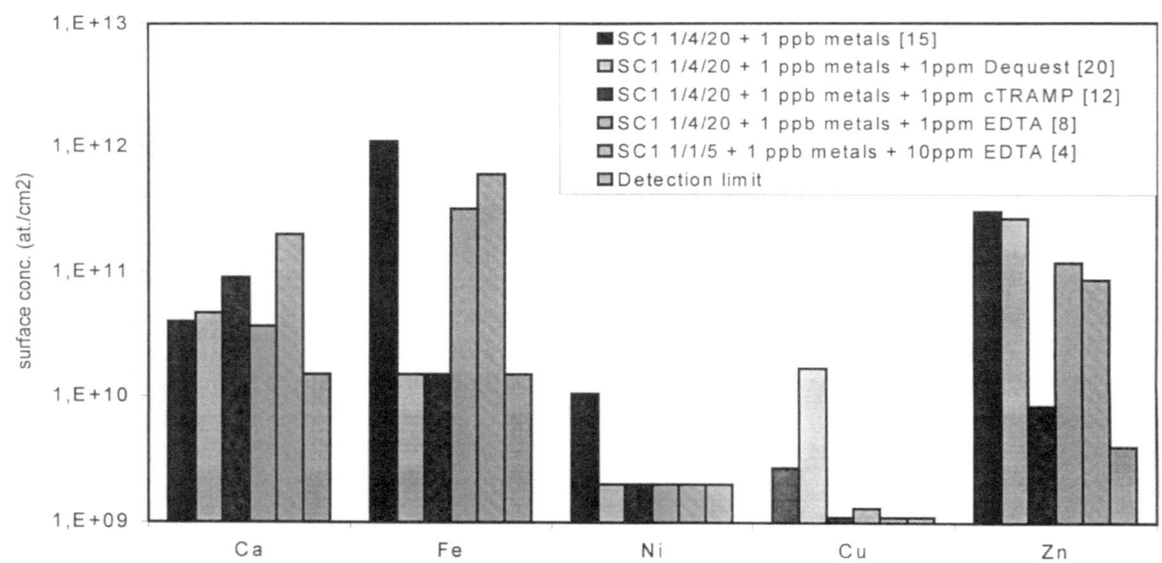

Fig. 4. Metal surface concentrations of SC-1 cleaned wafers with different chelating agents without intentional metal contamination.

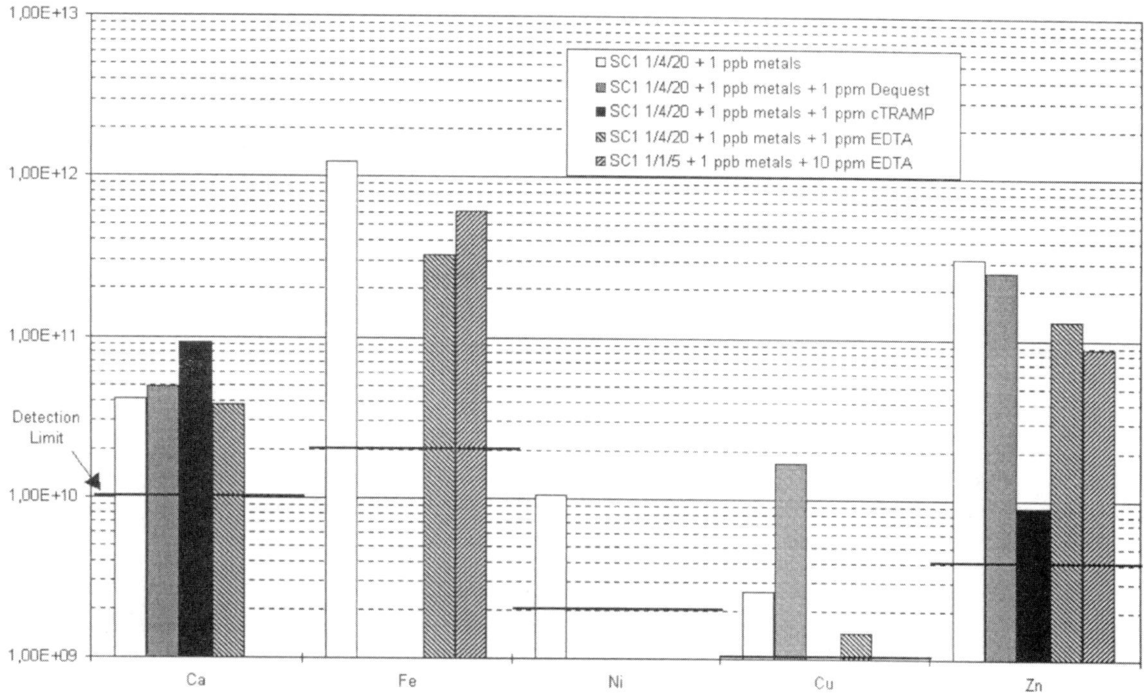

Fig. 5. Metal surface concentrations of SC-1 cleaned wafers with intentional metal contamination and different chelating agents added.

EDTA is not useful. Dequest 2060s is active with iron and nickel but surprisingly increases copper surface concentration.

5. Surface photo voltage (SPV) and ELYMAT measurements

Iron bulk concentrations calculated from SPV measurements of cleaning experiments (two wafers per experiment) without metal spiking are shown in Fig. 6.

In a pure SC-1 with a concentration ratio of 1/4/20 Fe bulk concentrations are fairly high. Dilution of ammonia and hydrogen peroxide (1/3/80) reduces the Fe bulk concentration. Addition of EDTA, Dequest and cTRAMP, respectively, lowers the Fe bulk concentration to 10^{10}–10^{11} atoms/cm^3.

SPV measurements are less sensitive than VPD–DSE–TXRF analysis. Therefore minority carrier life-time and diffusion length, respectively, were determined via ELYMAT photocurrent measurements (Fig. 7). These confirmed the results of metal surface analysis by VPD–DSE–TXRF.

Without intentional metal contamination pure SC-1 (1/4/20) still contains fairly high levels of residual metal contamination giving rise to poor photocurrents. Dilution to 1/3/80 and reduction of temperature to 50°C result in substantial increase of the photocurrent. Addition of EDTA leads to values of photocurrent comparable to those of references. Wafers cleaned in metal spiked SC-1 (1/4/20), with and without EDTA, and SC-1 (1/3/80) without EDTA exhibit extremely low (< 1 μA) photocurrents. Merely by addition of 1 ppm EDTA in SC-1 (1/3/80) photocurrents reach a level

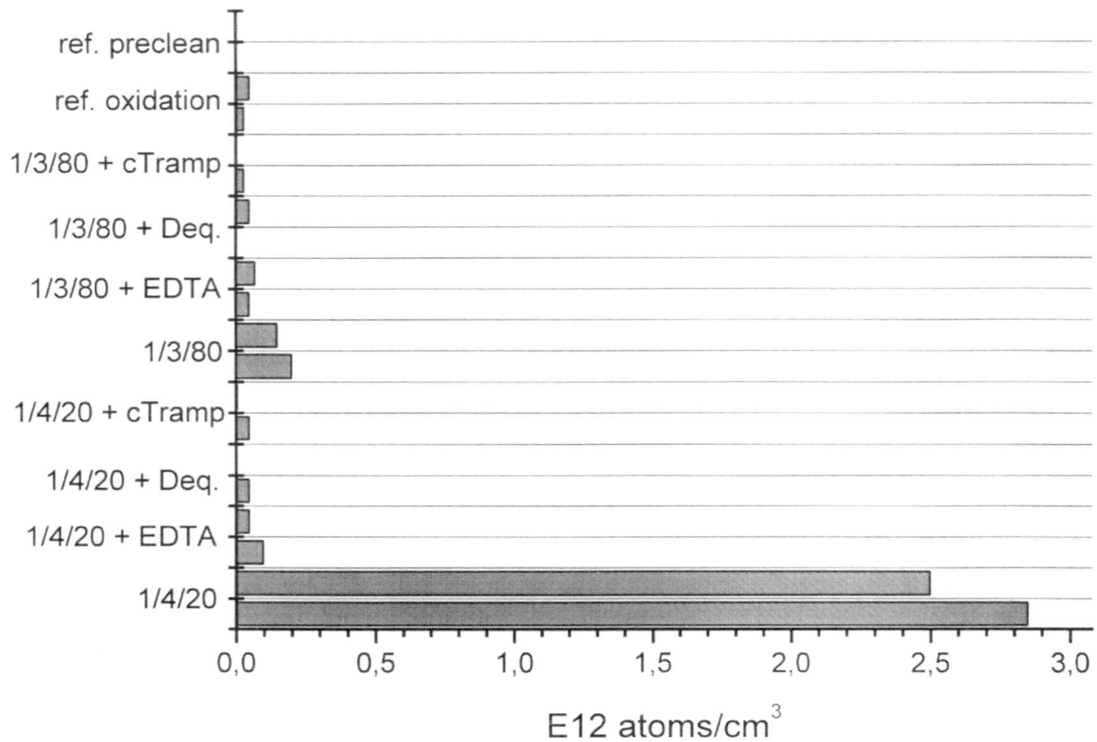

Fig. 6. Iron bulk concentrations from surface photo voltage (SPV).

comparable to that without intentional metal contamination. Addition of Dequest 2060s and cTRAMP, respectively, with and without metal spiking provides photocurrents comparable to the reference.

5.1. Hydrogen peroxide decomposition tests

In Figs. 8–10 the H_2O_2 concentration monitored via measurement of the oxygen partial pressure increase due to H_2O_2 decomposition is plotted as a function of bath age. Fig. 8 illustrates how addition of 1 ppb of various metal ions to pure SC-1 ('GB') influences the decomposition of hydrogen peroxide.

Fe is the most active metal ion in H_2O_2 decomposition followed by Cu, while Al, Ca, Ni and Zn play a negligible role in the decomposition process.

Fig. 9 exhibits that cTRAMP does not exert any detrimental effect while diammonium–EDTA decreases the bath lifetime due to free Fe-ions released by decomposition of Fe–EDTA-complexes. (The latter are originating from Fe contamination of the EDTA fabrication process.)

Fig. 10 demonstrates that, on the other hand, EDTA as free acid as well as its ammonia salt are able to slow down hydrogen peroxide decomposition in metal contaminated cleaning solutions. However, they are distinctly less effective than Dequest 2060s and cTRAMP.

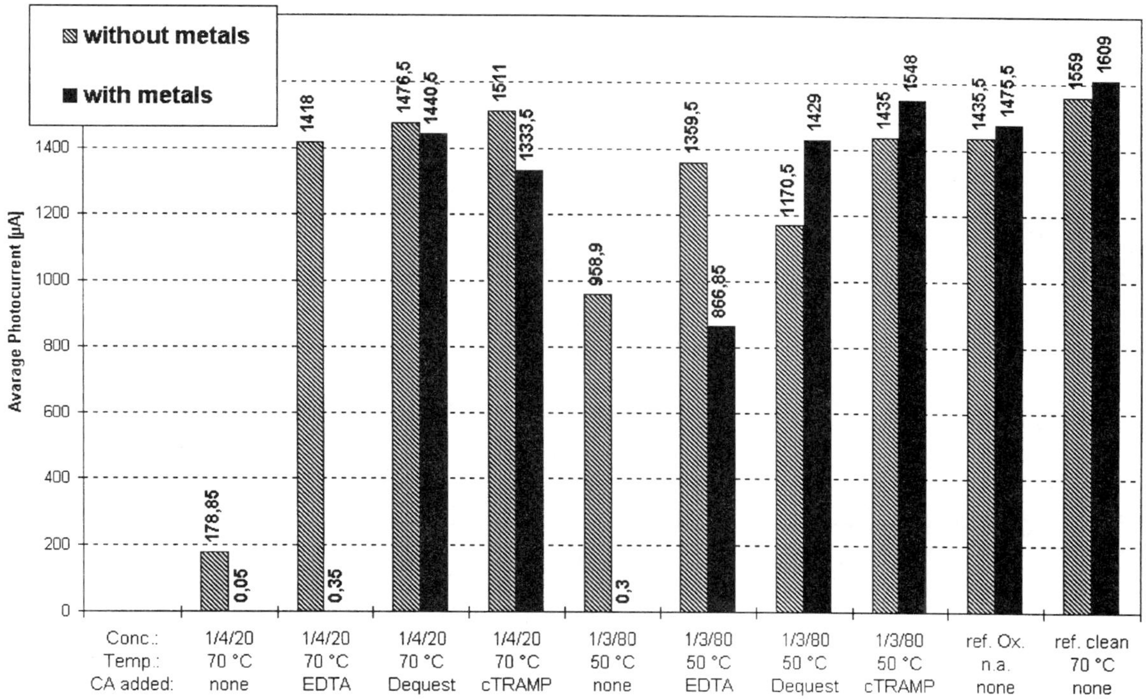

Fig. 7. Average photocurrent after rapid thermal annealing of SC-1 cleaned wafers with and without intentional metal contamination.

5.2. Particles (light point defects LPDs) and organic residues

According to LPD monitoring by TENCOR and CENSOR equipment no additional particles were counted due to CA addition to the cleaning solution. In TOF–SIMS studies various organic species covering the wafer surfaces were detected. However, no phosphorous containing species were found which should be expected if chelating agents with phosphonic acid groups (Dequest 2060s, cTRAMP) or their decomposition products are deposited on the surface. Therefore adsorption of Dequest 2060s and cTRAMP on the wafer surface is rather unlikely, the organics detected are primarily related to species from clean room environment, packaging and handling.

5.3. Gate oxide integrity (GOI) tests

The influence of cTRAMP on GOI has been measured on 15 nm gate oxide by ramped voltage stress (EBD) of 16 mm^2 capacitors and constant current stress (QBD) on 0.126 mm^2 capacitors at 0.5 A/cm^2 on CZ and Epi wafers with and without metal spiking of the SC-1 cleaning solution used before oxidation (Fig. 11). Addition of cTRAMP to intentionally contaminated SC-1 solutions completely suppresses the devastating effect of the metals in SC-1: No decrease in QBD yield could be observed, the yield is the same as in pure SC-1 chemistries and as the standard clean reference

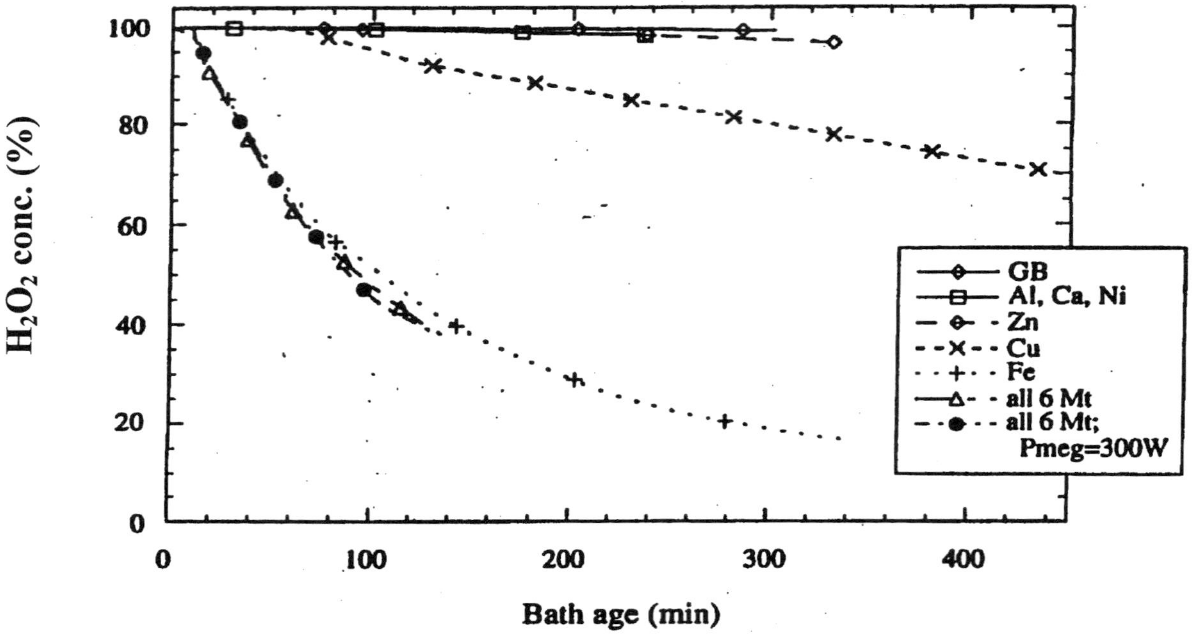

Fig. 8. Influence of various metal ions on the decomposition of hydrogen peroxide. (Reference: Gigabit grade SC-1 ('GB')).

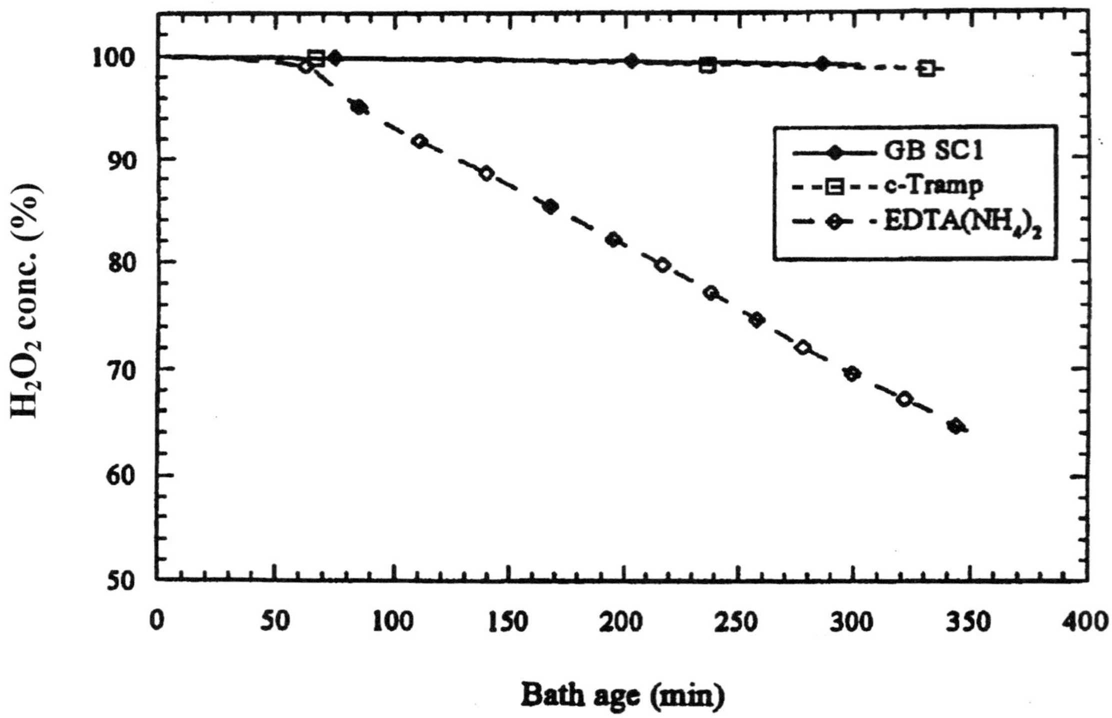

Fig. 9. Effect of cTRAMP and EDTA on the H_2O_2 decomposition compared to pure SC-1. (GB SC-1).

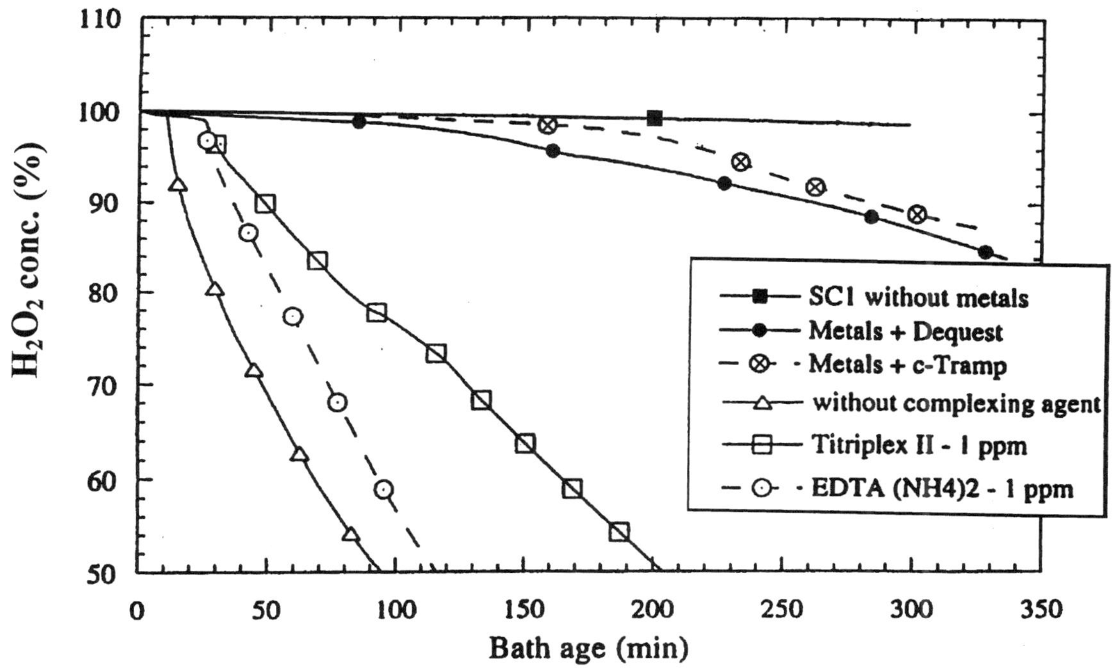

Fig. 10. Comparison of the capability of EDTA, Titriplex V™, cTRAMP and Dequest 2060s to stabilize H_2O_2 in SC-1 spiked with metals (1 w-ppb Al, Ca, Cu, Fe, Ni, Zn). Reference: pure SC-1.

(Fig. 11). Spiking of cTRAMP and chemicals during 3 hours bath lifetime of contaminated SC-1 still prevents metal precipitation on the wafer surface.

The EBD was not sensitive to metal surface contamination prior to oxidation due to the thickness (15 nm) of the gate oxide.

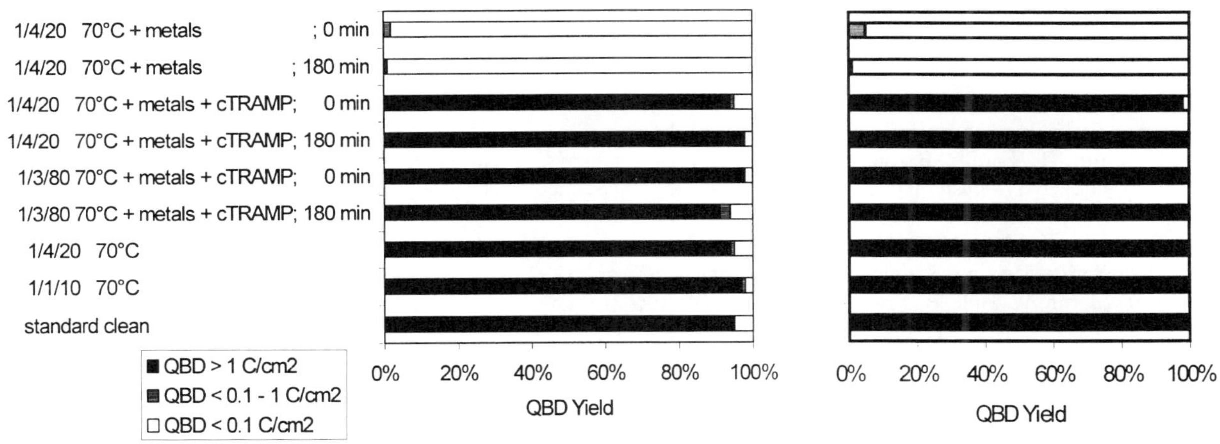

Fig. 11. GOI results after wafer cleaning under various conditions.

6. Summary and conclusions

The heterocyclic compound cTRAMP represents a new class of chelating agents for wafer cleaning in semiconductor manufacturing. Addition of cTRAMP in the ppm range to a SC-1 solution enables the replacement of the entire RCA-sequence by a single step cleaning. The ability of SC-1 solutions to remove particles while controlling the amount of Fe, Zn and Cu metal ions added to the wafer surface has been demonstrated for conventional (1:4:20, 70°C) as well as diluted (1:3:80) mixtures and reduced temperatures (50°C).

Minority carrier lifetimes, diffusion length and gate oxide integrity, respectively, are improved. No deposition of the chelating agent is detected by TOF–SIMS onto the wafer surface, no particles are added and roughness is not increased. Moreover, decomposition of hydrogen peroxide is reduced enabling longer bath lifetime. This excellent combination of properties makes cTRAMP an attractive candidate as additive in SC-1 to provide single step cleaning with prolonged bath life and high tolerance to unintentional and cross metal contamination.

Acknowledgements

The technical and scientific support by Dr. Albrecht Fester, Harald Herrmann and Claudia Rittmeyer and the assistence by Erika Schaffert in the preparation of the manuscript is gratefully acknowledged.

References

[1] W. Bergholz, G. Zoth, F. Gelsdorf, B.O. Kolbesen, Metal Contamination in ULSI Technology, Proceedings of the Second Symposium on Defects in Silicon II, The Electrochemical Society, Proc. Vol. 91–9, p. 21
[2] B.O. Kolbesen, Defect aspects of advanced device technologies, in: S. Coffa, F. Priolo, E. Rimini, J.M. Poate (Eds.), Erice/Italy, July 1991, Proceedings of the International School on Crucial Issues in Semiconductor Materials and Processing Technologies, NATO ASI Series, Vol. 222, Kluwer Academic Publishers, Dordrecht, The Netherlands, 1991, p. 3.
[3] W. Kern, David A. Puotinen, RCA Review 31 (6) (1970) 187.
[4] H.F. Schmidt, M. Meuris, P.W. Mertens, A.L.P. Rotondaro, M. Heyns, T.Q. Hurd, Z. Hatcher, Jpn. J. Appl. Phys. 34 (1995) 727.
[5] M. Meuris, P.W. Mertens, A. Opdebeeck, H.F. Schmidt, M. Depas, G. Vereecke, M.M. Heyns, A. Philipossian, Solid State Technology, July 1995, p. 109
[6] F. Tardif, T. Lardin, B. Sandrier, P. Boelen, R. Matthews, I. Kashkoush, R. Novak, in: Proceedings of Fifth International Symposium on Cleaning Technology in Semiconductor Device Manufacturing, The Electrochemical Society, Pennington, NJ, 1998, p. 15, Proc. Vol. 97–35.
[7] H. Morinaga, M. Aoki, T. Maeda, M. Fujisue, H. Tanaka, M. Toyoda, in: Proceedings of Material Research Society 1997 Spring Meeting, Symposium P, San Francisco, April 1997.
[8] A.R. Martin, W. Hub, B.O. Kolbesen, Patent DE 196 31 363
[9] R.D. Hancock, A.E. Martell, Chem. Rev. 1989, 89 (1875)
[10] J. Inczedy (Ed.), Analytical Applications of Complex Equilibria, Ellis Horwood Ltd, Chichester, UK, 1976, p. 63.
[11] H.F. Schmidt, M. Meuris, P.W. Mertens, A.L.P. Rotondaro, M. Heyns, T.Q. Hurd, Z. Hatcher, in: Proc. 2nd Int. Symp. Ultra Clean Processing of Silicon Surfaces, Acco, Leuven, Belgium, 1994, p. 259.

ELSEVIER

Microelectronic Engineering 45 (1999) 209–223

MICROELECTRONIC ENGINEERING

Mechanical strength of 300 mm diameter silicon wafers at high temperatures: modeling and simulation

A. Fischer[a,*], Th. Grabolla[a], H. Richter[a], G. Obermeier[b], P. Krottenthaler[b], R. Wahlich[b]

[a]*Institute for Semiconductor Physics, Walter-Korsing-Str. 2, 15230 Frankfurt (Oder), Germany*
[b]*Wacker Siltronic AG, Johannes-Hess-Str. 24, 84489 Burghausen, Germany*

Abstract

Under gravitational and thermal constraints of IC process technology, 300 mm diameter silicon wafers can partly relax via slip dislocation generation and propagation, degrading the electrical characteristics of the leading edge device. We present a force balance model to describe the strain relaxation in large wafer diameter, which includes heat transfer effects and the criterion for yielding under a plane stress state. The material attributes, e.g. oxygen and its state of aggregation, are taken into account. While the plastic deformation of silicon wafers caused by thermal stresses at high temperatures can be controlled by process design, the control of plastic deformation due to gravitational forces may be accomplished by equipment design. This system approach allows calculation of wafer mechanics and ramp rate profiles for an arbitrary high-temperature process. The quantitative theory proposed here provides guidance for computer simulation to configure stable slip-free wafer process flow under mechanical and thermal loads. Applications include high speed simulations for use in 'what if' experiments or initial simulations of large scale experimental sequences. The simulator developed can also be used by IC manufacturers to determine optimum wafer throughput and cycle times in front-end device processes.

Keywords: Large silicon wafers; Mechanical strength; Modeling and simulation

1. Introduction

Conversion to 300 mm diameter silicon wafers driven by increased chip sizes is necessary to achieve the required economies of scale for high volume IC manufacturing. There are a number of engineering challenges which face the silicon wafer and IC equipment communities in bringing 300 mm wafer utilization to a production-level maturity. Slip-free high temperature processing of large silicon wafer diameters is currently a primary challenge. Plastic deformation of silicon wafers caused by thermal stresses at high temperatures can be prevented by appropriate process design. Control of plastic deformation due to gravitational forces and film induced stresses often requires appropriate equipment design (see Fig. 1).

*Corresponding author. Tel.: +49-335-56250; fax: +49-335-562-5300.
E-mail address: fischer@ihp-ffo.de (A. Fischer)

0167-9317/99/$ – see front matter
PII: S0167-9317(99)00155-0

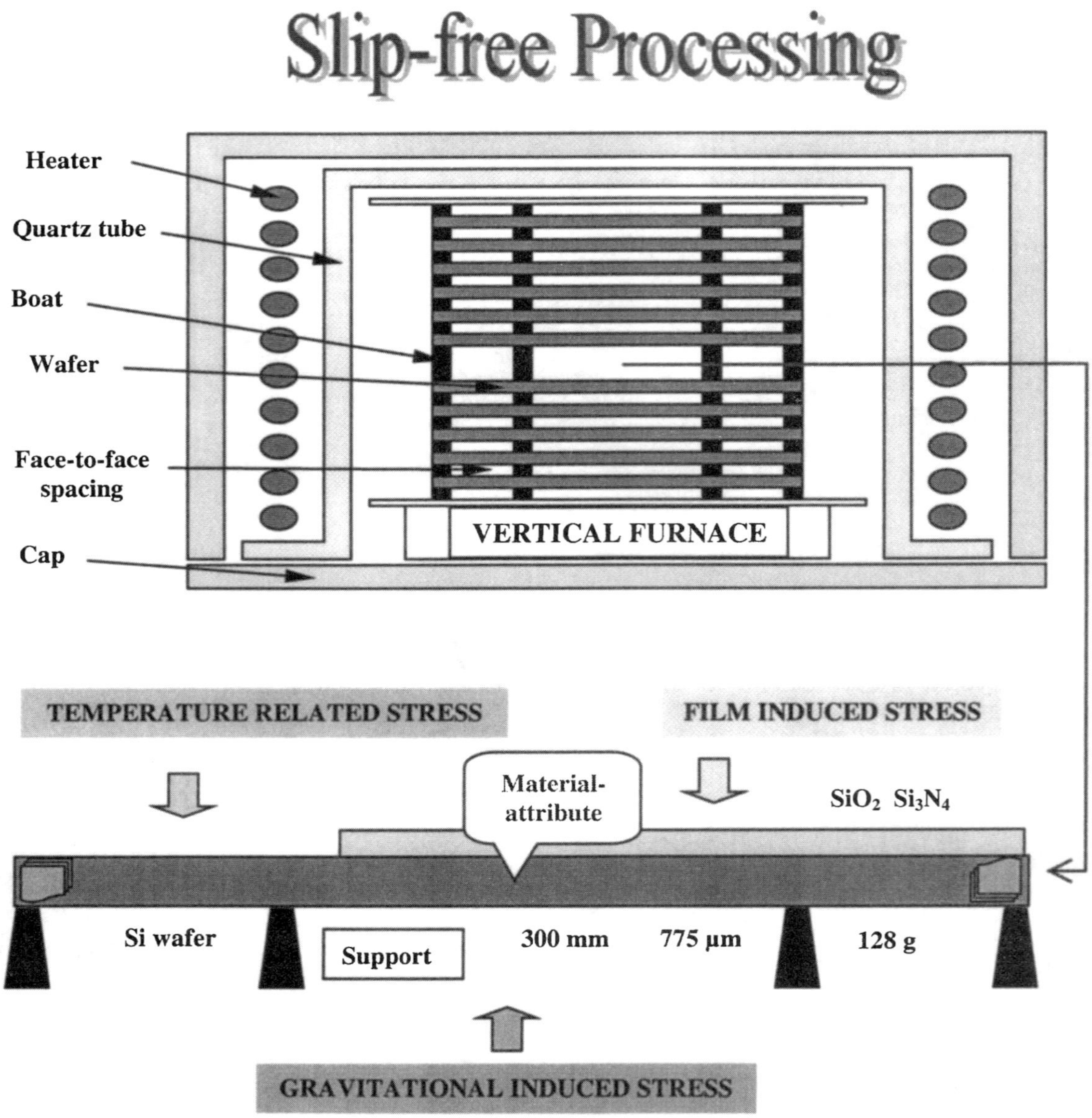

Fig. 1. Schematic illustration of large diameter silicon wafers under mechanical and thermal constraints of process technology lying in a vertical batch-type furnace.

In this paper, we present a force balance approach to describe the onset of plastic strain relaxation in large diameter silicon wafers subjected to mechanical and transient thermal loads. With a view to computer simulation, the treatment of the biaxial state of stress and shear stress components on the slip planes has been developed in a simple analytical form, valid in linear elasticity. The present discussion is limited to an isotropic approximation for the crystal silicon. To determine the high temperature processing windows in the IC fabrication without the need for experimental backup, we adequately model the complex phenomenon of bending, compressive, and thermal stresses in batch wafers during various processing steps. The present force balance model has been calibrated using the experimental results and the limits of furnace power. We show how the process simulator developed and experimentally verified calculates slip-free ramp up and down regimes for different wafer support configurations. We predict behaviors that have significant impact on the understanding of high temperature treatment of large silicon wafers and shed light on some remarkable theoretical results. The simulator developed can also be used by IC manufacturers to determine optimum wafer throughput and cycle times in front-end device processes.

2. Force balance model for strain relaxation in silicon wafers

In thermodynamic equilibrium, slip dislocations appear at the slip planes of a strained silicon wafer when the resolved shear stress due to mechanical and thermal constraints of process technology reaches the upper yield stress of the covalently bonded material. The onset of yielding is a measure of the mechanical strength of the silicon wafers and depends on temperature and specific material properties, e.g. content of oxygen and its state of aggregation. The criterion for yielding involves only components of shear stress.

The procedure to determine the process conditions which prevent plastic deformation for a particular experimental configuration involves writing the equations for the resolved shear stress components that act on the slip systems as a consequence of externally applied loads and the equation for the upper yield stress τ_{uy} as a function of the temperature T and the precipitated oxygen content $\Delta[O_i]$. In the general case of plane stress distribution in a thin wafer, the shear stress components are the gravitational induced bending stress τ_b as a function of the support configuration and the wafer weight, the gravitational induced compressive stress τ_c as a function of the support area and the wafer weight, and the temperature related stress τ_T in terms of a resolved shear stress of radial temperature distribution in the wafer, characterized by a difference between wafer edge and center temperatures ΔT. Note that the stresses and strains caused by such external sources of stress, i.e. body forces, can be superposed [1]. Combining the four stress terms, we easily obtain an expression for the equilibrium state of stress in the wafer to prevent plastic deformation, to wit

$$\tau_{uy} - \tau_b - \tau_c - \tau_T = 0 \qquad (1)$$

The components of this crucial stability criterion are discussed in the following sections.

3. Gravitational induced stresses

Three hundred mm silicon wafers horizontally stacked and ring- or point-like supported in a vertical-type furnace bend downward due to their own weight. In the latter case, additional attention should be paid to the compressive contact stress near the points of application of the load. In linear elasticity, the gravitational induced bending stress is linearly distributed along the wafer thickness, and reaches its maximum in the contact region on the upper and lower wafer surfaces in either tensile or compressive condition. This stress was evaluated by Fukuda [2] for three- and four-point supports located symmetrically at the wafer perimeter. By taking some correction terms in his analysis, we may obtain solutions of bending problems for various symmetrical configurations of application of the load. In the general symmetrical case, the maximum bending stress can then be represented in terms of a resolved shear stress by the following series:

$$\tau_b = \cos\lambda\cos\phi\,\frac{3\rho g(R-\Delta R)^2}{4t}\left[1-\nu-\frac{16(1+\nu)}{3+\nu}\frac{1}{n}\sum_{k=1}^{360n^{-1}}\frac{1}{k}\right], \qquad (2)$$

where $\cos\lambda\cos\phi$ is the Schmid factor (=0.408), ρ is the silicon mass density, g is the gravitational acceleration, R is the wafer radius, ΔR is the difference between wafer radius and support radius, n is the number of support points, and t is the wafer thickness. Note that, for a ring-like support, we have to choose $n=360$. Table 1 lists the computed maximum values of bending stress τ_b for 200 mm and 300 mm silicon wafers held in various symmetrical support configurations. For direct comparison, calculated stress values obtained by several authors are also shown in Table 1. The agreement between the numerical values of bending stress for same support conditions is evident.

Compressive contact stress is also a form of gravitational induced stress since contact forces derive from the wafer weight. In this case the body-force problem reduces to an edge-load problem. This stress was calculated by Filon [5] for a long strip when the thickness of the strip is small in comparison with the length l. Letting in Filon's formula $R=l$ and $q=\rho g t$, we can also find solutions for a thin circular plate. These solutions are satisfactory only if the distance between the support points remains large in comparison with the wafer thickness. We get the following expression for the compressive contact stress in the wafer:

Table 1
Maximum gravitational resolved shear stress τ_b (MPa) for different wafer supports

Wafer diameter	Support design	This work	Fukuda [2] (Fujitsu)	Huff [3] (Sematech)	Griffith [4] (Sandia)
200 mm	Ring at the edge	−0.07		−0.07	
	Ring at 0.7 R	−0.03		−0.03	
	Symmetrical 3-point edge	−1.01			−1.02
	Symmetrical inner (0.7 R) 3-point	−0.49			−0.46
	Symmetrical 4-point edge	−0.70			−0.70
	Symmetrical inner (0.7 R) 4-point	−0.34			−0.31
300 mm	Ring at the edge	−0.14			
	Ring at 0.7 R	−0.07			
	Symmetrical 3-point edge	−2.12	−1.60		
	Symmetrical inner (0.7 R) 3-point	−1.04			
	Symmetrical 4-point edge	−1.46	−1.10		
	Symmetrical inner (0.7 R) 4-point	−0.72			

$$\tau_c = -\cos\lambda\cos\phi\, q\left[\frac{a}{R} + \frac{4}{\pi}\sum_{m=1}^{\infty}\frac{\sin\left(\frac{m\pi a}{R}\right)}{m}\,\frac{\frac{m\pi t}{2R}\cosh\left(\frac{m\pi t}{2R}\right) + \sinh\left(\frac{m\pi t}{2R}\right)}{\sinh\left(\frac{m\pi t}{R}\right) + \frac{m\pi t}{R}}\cos\left(\frac{m\pi x}{R}\right)\right], \tag{3}$$

with the uniformly distributed load of intensity q over the contact area na^2

$$q = \frac{\rho g \pi R^2 t}{na^2}$$

and an approximated value of $a \approx \sqrt{10\,000\ \mu\text{m} \times 50\ \mu\text{m}}$. Here x is the lateral distance from the support point. For the case of four-point supports, the result of this calculation is shown in Fig. 2. It will be seen that the local compressive stress in the vicinity of the point of application of the load due to the 300 mm wafer weight decreases very rapidly with increase of distance x from the support point, and at a value $x \approx t$, τ_c becomes zero and is then replaced by weak tension. The maximum stress

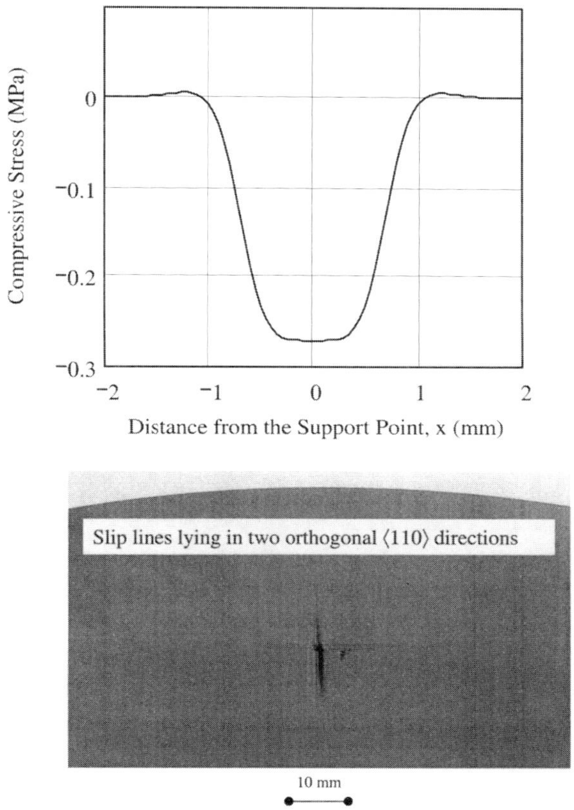

Fig. 2. Local stress distribution near the point of application of the load due to the 300 mm wafer weight for the case of four-point support (above) and slip line network built in the vicinity of the contact points with the wafer support after thermal treatment (Lang topograph image, below).

values takes place in the center of the contact region. To illustrate the application of this method of compressive stress calculation in silicon wafers, let us consider the case shown in Fig. 2, below. The density of distribution of dark slip lines indicates high stresses near the point of application of the load. The number of slip lines crossing a cross section diminishes as the distance of the cross section from the middle of the contact region increases. This is due to rapid decrease in gravitational induced compressive stress. These results are of high practical importance. To reduce the contact stress, the shape of the support points should have a cylindrical surface with a radius of curvature in the order of the support radius. Appropriate design of support points involves industrial techniques to be developed.

4. Temperature related stress

In batch-type high temperature furnaces, as wafers are stacked with less face-to-face spacing, the radiation heat flux from the hot-wall of the furnace to the wafer surface is not uniform across the wafer. The transient temperature profile across a wafer can be expressed in a parabolic function of the radial position r [6,7]:

$$T = T_{\mathrm{Center}} + \Delta T\left(\frac{r}{R}\right)^2,$$

where T_{Center} is the temperature corresponding to the wafer center, and ΔT is the temperature difference between the edge and center of the wafer. During heating, the temperature gradient produces a tensile stress in the central region and a dominant tangential component σ_θ of compressive stress in the outer region. The radial component σ_r, which is everywhere tensile, decreases to zero at the wafer edge, but is equal to σ_θ at the wafer center. $\sigma_\theta(R)$ at the edge is always much larger than $\sigma_\theta(0)$ at the center. During cooling, the senses of these stress components are reversed. Once the temperature distribution is calculated, the distribution of stress σ_θ and σ_r can be found easily for axisymmetric plane stress problems [1,8,9]. Because of axisymmetry, the shear component $\sigma_{\theta r}$ is zero. Then, the maximum resolved shear stress τ_T that acts on the slip systems $\{111\}\langle 110\rangle$ in $\{100\}$ silicon wafers having radial temperature gradients is given by [8]

$$\tau_T = \cos\lambda \cos\phi \frac{E\alpha(T)}{4} \Delta T[1 - (2 + \sin 2\theta + \cos 2\theta)], \quad (4)$$

where E is the Young modulus, $\alpha(T)$ is the linear thermal expansion coefficient of silicon, and θ is the angular position of the primary slip systems on the wafer.

Fig. 3 shows an example of thermal stress in a silicon wafer during heating. This in-plane stress distribution produced by non-uniform heating or cooling results in circumferential buckling which can result in a permanently warped wafer if the temperature is sufficiently high to allow plastic flow. Slip patterns thus obtained are also shown in Fig. 3, below. In this case, the maximum resolved shear stress due to temperature related stress exceeds the upper yield stress of silicon crystal and the wafer strain may give way to slip dislocations leading ultimately to wafer warpage.

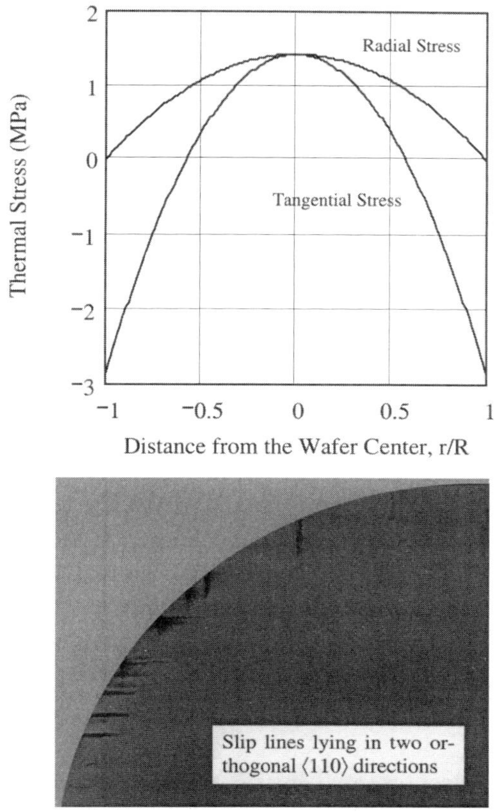

Fig. 3. Thermal plane stress distribution over the wafer for an axisymmetric difference between wafer edge and center temperature $\Delta T = 10$ K in heating environment. The stress components are plotted versus the normalized distance from the wafer center r/R (above). Lang topograph image of slip line propagation caused by thermal stresses due to non-uniform heating or cooling of the wafers up to or down from the process temperature (below).

5. Upper yield stress of silicon crystals

When growing Czochralski silicon crystals, it is inevitable that oxygen and carbon will contaminate the silicon, and these impurities greatly influence the mechanical strength of silicon. At high temperatures, dissolved oxygen atoms and small oxygen clusters congregate at dislocations, locking them firmly and preventing the dislocations from multiplying. These oxygen atoms thus improve the mechanical strength of silicon crystals at high temperatures. Carbon atoms dissolved in Czochralski silicon crystals behave as nuclear centers for oxygen precipitation. Dislocation loops, punched out from such oxygen precipitates formed at high temperatures, decrease the upper yield stress because the punched-out dislocations act as slip dislocation sources [10,11]. It should be noted that nitrogen was also found to be very effective in enhancing oxygen precipitation in low oxygen wafers under various heat treatments. The interaction between dopants and point defects also affects oxygen precipitation. Heavily doped p-type wafers are much easier to precipitate than heavily doped n-type silicon wafers [12]. On the other hand, p^+ silicon substrate are expected to pin on the dislocation more strongly than dissolved oxygen atoms [13]. Although dislocation pinning stress in p^+ wafers has not been understood well, this must be accounted for in process design and modeling.

The major factor affecting oxygen precipitation in silicon wafers is the amount of supersaturated oxygen at the heat treatment temperature. For a given thermal cycle, the dependence of the change in interstitial oxygen concentration, $\Delta[O_i]$ (dissolved oxygen loss ≅ amount of precipitation), upon the initial oxygen concentration, $[O_i]_0$, is described by a specific curve having a predictable functional dependence [12]. Finding the specific curve is the first step in finding the optimum range of $\Delta[O_i]$ that will provide adequate gettering ability without producing slip dislocations and wafer warpage. A specific curve $\Delta[O_i]$ versus $[O_i]_0$ is shown in Fig. 4.

The upper yield stress of the covalently bonded crystal such as silicon decreases with increasing temperature T, dislocation densities, precipitated oxygen content $\Delta[O_i]$, and with slower strain rates. The semiempirical equations for the upper yield stress τ_{uy} of dislocation-free silicon according to the experimental values given by Leroy and Plougonven [14] and Schrems et al. [15] become

$$\tau_{uy} = L_i \tau_P \exp\left[2.4\left(1 - \frac{T}{273\ \mathrm{K}}\right)\right] \tag{5a}$$

and

$$\tau_{uy} = S_i \dot{\epsilon} \exp\left(\frac{0.95\ \mathrm{eV}}{kT}\right), \tag{5b}$$

respectively. Here τ_P is the Peierls stress, $\dot{\epsilon}$ is the strain rate $(=9\times10^{-7}\ \mathrm{s}^{-1})$, L_i and S_i are empirical constants depending on $\Delta[O_i]$, and k is the Boltzmann constant. Experimental results of the silicon yield point for both data sets are plotted in Fig. 5. It is seen that, in the temperature range 700°C–1100°C, the values of τ_{uy} represented by Eq. (5a) are much larger than the values calculated using Eq. (5b). Further yield stress experiments would be required to find appropriate key data for 300 mm silicon material. For our further investigations, we will employ the upper yield stress values given in Ref. [14].

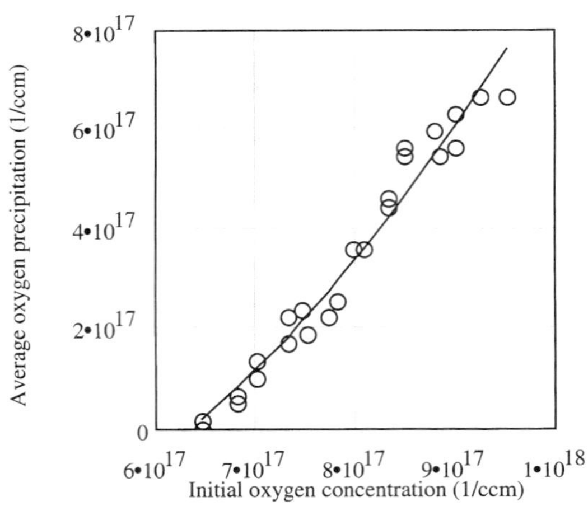

Fig. 4. Polynomial regression curve (solid line) and experimental (circles) average oxygen precipitation $\Delta[O_i]$ as a function of initial oxygen concentration $[O_i]_0$ for a CMOS process simulation.

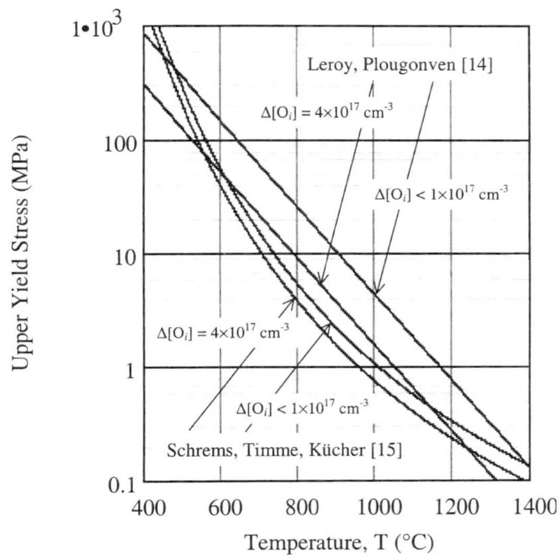

Fig. 5. Upper yield stress data of dislocation-free Czochralski silicon crystals plotted against the temperature for different precipitated oxygen contents $\Delta[O_i]$.

6. Maximum allowable ramp rate

As shown above, the upper yield stress at 1200°C decreases to within a few tenths of a MPa, which is very close to the stress caused only by gravitational induced compressive stress τ_c near the point of application of the load due to the 300 mm wafer weight for the case of four-point supports. From this we conclude that a four-point support of this configuration is problematic for processing at 1200°C because of the occurrence of slip as shown in Fig. 2. However, with an increase in support points for wafers, e.g. by using a six-point support or even a ring-like support, the gravitational induced stresses τ_c and τ_b can be decreased drastically, so that thermal stresses τ_T can be applied for ramping up to 1200°C. Notice that similar problems of 300 mm wafer warpage have been reported in Ref. [16], where extremely high annealing temperatures and a symmetrical inner three-point support had been used.

For a complete description of the equilibrium state of stress in the wafer to prevent its plastic flow during ramping, we should now consider the magnitude of the temperature related stress τ_T that can act on the slip systems. Rearranging Eq. (1) we get an expression for this critical resolved shear stress. The corresponding allowable temperature difference ΔT, satisfying Eq. (1), can then be taken in the form of

$$\Delta T(T) \leq \frac{\tau_{uy} - \tau_b - \tau_c}{\cos\lambda \dfrac{E\alpha(T)}{4}[1-(2+\sin 2\theta + \cos 2\theta)]}, \tag{6}$$

in which the functions τ_{uy}, τ_b, and τ_c are given by Eqs. (5a), (2), and (3), respectively. This relationship is plotted in Fig. 6 for different oxygen concentrations as an important material property.

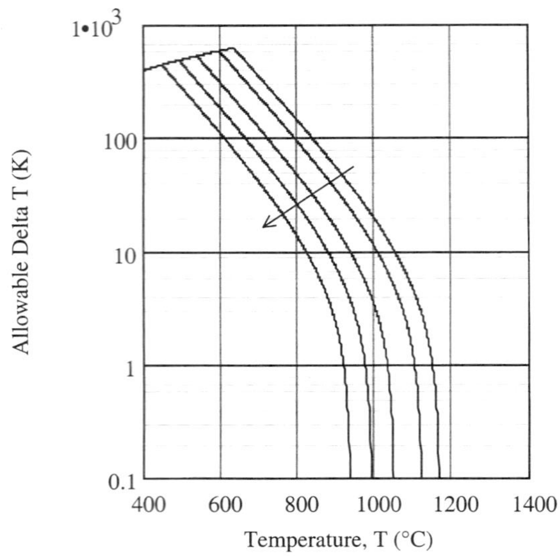

Fig. 6. Allowable temperature T and temperature difference ΔT as a function of T for different precipitated oxygen contents $\Delta[O_i]<1$, 1, 4, 7, 10×10^{17} cm^{-3} (in the direction of the arrow), accounting for gravitational stress in a 300 mm wafer on a 0.7 R inner four-point support.

In the absence of heat conduction at intermediate and large wafer spacings in the batch row, the relationship between maximum allowable ramp rate V and allowable temperature difference ΔT is well approximated by a radiation term [7,17–19]

$$V(T)=\frac{8f_c\varepsilon\sigma}{\rho ct}\left[T^4-(T-\Delta T(T))^4\right] \tag{7}$$

with the geometrical factor f_c for cases with contribution from multiple reflections k

$$f_c=1-\frac{\varepsilon}{2}\sum_{k=0}^{\infty}(1-\varepsilon)^k\left[1+\frac{\frac{R^2}{w^2}-(k+1)^2}{\frac{R^2}{w^2}+(k+1)^2}\right].$$

Here, ε is the radiation emissivity of the wafer surface, σ is the Stephan–Boltzmann constant, c is the specific heat of the wafer material, and w is the face-to-face wafer spacing in the boat. A plot of V versus T is shown in Fig. 7. Inspection of Figs. 6 and 7 shows some interesting relationships. For $\tau_{uy}(T,\Delta[O_i])\cong\tau_b+\tau_c$, the temperature related stress τ_T vanishes and the quantity ΔT decreases to zero. For $\Delta[O_i]<1\times10^{17}$ cm^{-3} and a 300 mm wafer resting on a 0.7 R inner four-point support, a maximum process temperature of about 1150°C results. It should be noted that, using upper yield stress data given by Eq. (5b), this maximum temperature becomes approximately 1000°C for the same support configuration.

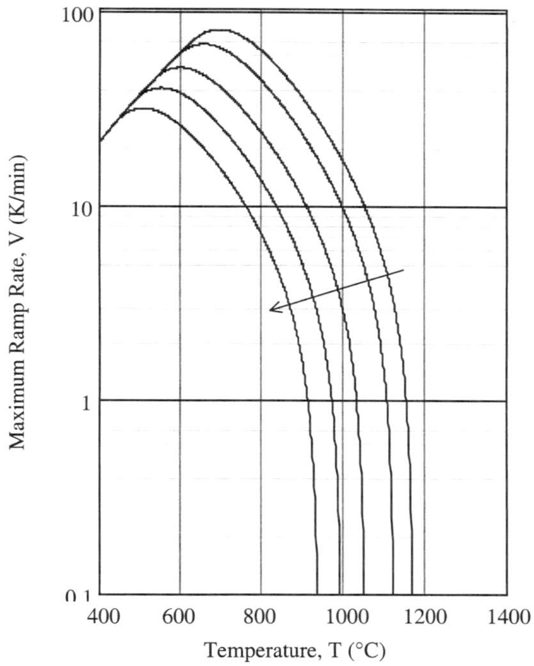

Fig. 7. Maximum allowable ramp rate V as a function of temperature T for different precipitated oxygen contents $\Delta[O_i]<1$, 1, 4, 7, 10×10^{17} cm^{-3} (in the direction of the arrow), accounting for gravitational stress in a 300 mm wafer on a 0.7 R inner four-point support and a face-to-face wafer spacing in the boat of 8 mm.

7. Simulation of thermal batch processing for large silicon wafer diameters

So far we have considered the mechanical and thermal constraints of process technology and the upper yield stress problems for large silicon wafer diameters. The force balance model described in the previous sections has been calibrated using the experimental results and the limits of furnace power. We thus arrive at the following computer simulation to configure slip-free ramp rate profiles for an arbitrary high temperature process.

A complete input/output deck for simulation of ramp processes in a batch-type fast ramp furnace including a program run for 300 mm wafers is shown in Fig. 8. The program code computes gravitational and thermal stresses in silicon wafers and uses these results to determine the total ramp duration, push–pull temperature, and the ramp rate profile for a given process temperature. Executing time is typically about 60 seconds on a Pentium-II processor operating at 300 MHz. Bending and contact stresses are first calculated based on the user's input of wafer and boat geometry. The difference between the upper yield stress and the gravitational induced stresses is then used to determine the allowable temperature variation across the wafer. Finally, a model of radial heat transfer in a batch furnace is used to compute the maximum allowable heating or cooling rate as a function of the allowable temperature difference between the edge and center of the wafer and the user's inputs of specific material attributes, wafer spacing, and maximum available heating and cooling capability of the furnace [20].

Thermal Batch Processing for Large Silicon Wafer Diameter:

Input data Si material (0 - 4) ii ≡ 0 Wafer radius R ≡ 150·mm Wafer thickness t ≡ 775·μm
Support radius (0.5R - R) Rs ≡ 105·mm Support points (3 - 360) Ps ≡ 4 Face-to-face spacing w ≡ 8·mm
Surface emissivity (0.5 - 0.9) $\varepsilon \equiv 0.65$ Process temperature (min800°C - max1200°C) TProc ≡ 1050·°C
Ramp up no. (0.3 - 1.3) RUN ≡ 0.6 Ramp up control RUC ≡ 30·K·min^{-1} Ramp up steps RUS ≡ 10
Ramp down no. (0.3 - 1.3) RDN ≡ 0.6 Ramp down control RDC ≡ 10·K·min^{-1} Ramp down steps RDS ≡ 10

Output data Total ramp duration DRamp = 78 •min Push-pull temperature TPP = 630 •°C
Push duration DPush = 25 •min Pull duration DPull = 7 •min Ramp rate profile for batch processes:

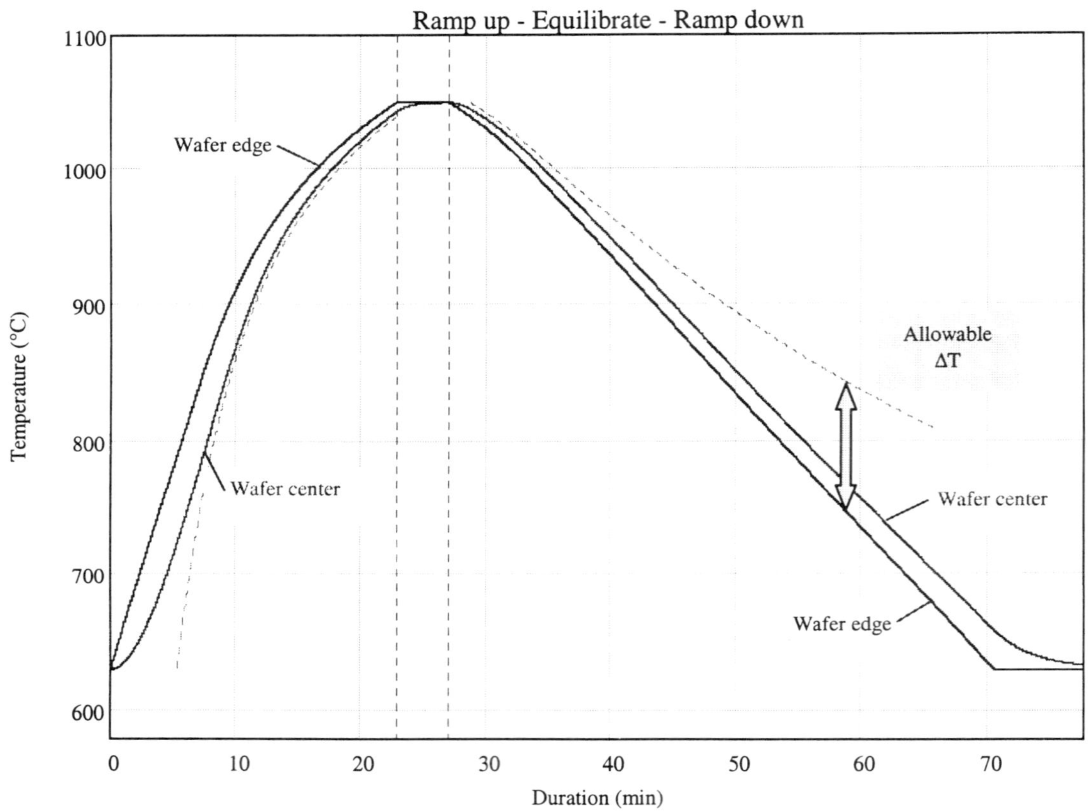

Fig. 8. Input/output table of the ramp simulator for 300 mm silicon wafers. The allowable radial difference between wafer edge and center temperatures ΔT is indicated. The ramp up rate is limited by furnace heater power (e.g. RUC = 30 K/min) and allowable ΔT. The ramp down rate is limited by furnace air cooling capacity (e.g. RDC = 10 K/min) and allowable ΔT.

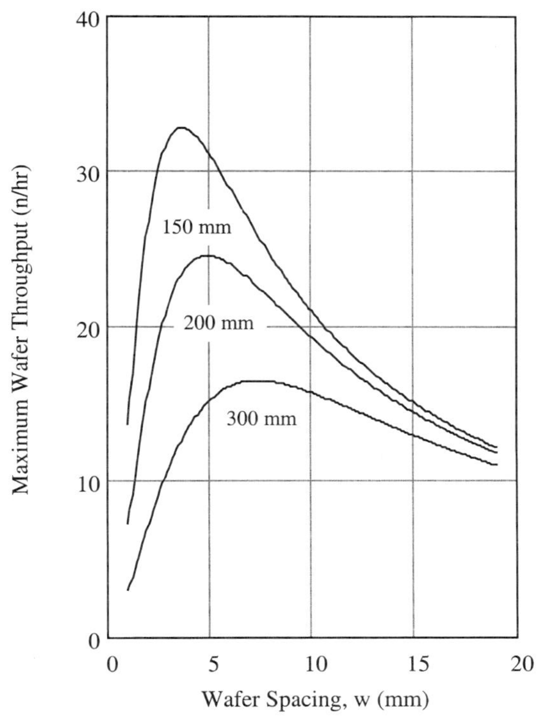

Fig. 9. Maximum wafer throughput per hour for a 2 h/1050°C high temperature process as a function of the face-to-face wafer spacing *w* for different wafer diameters and an effective furnace height of 500 mm.

The simulator developed can also be used by IC manufacturers to determine maximum wafer and areal throughput per hour in front-end device processes. For a standard 300 mm wafer process, Figs. 9 and 10 show throughput data calculated using the upper yield stress values given in Ref. [14]. For direct comparison, the relationships between throughput and wafer spacing in the boat for 200 mm and 150 mm wafers are also shown. It can be seen, although the wafer throughput decreases with increasing wafer diameter, the areal throughput increases dramatically. The optimum wafer spacings for 300 mm, 200 mm, and 150 mm silicon wafers are about 8 mm, 5 mm, and 4 mm, respectively.

8. Summary and conclusions

We have endeavored to describe the onset of plastic strain relaxation in large diameter silicon wafers in a consistent manner by relating the effect of gravitational and thermal constraints of IC process technology on the mechanical strength of the wafers. The present system of equations is generally sufficient to quantitatively determine the high temperature processing windows in the IC fabrication without the need for experimental backup. Although further yield stress experiments would be required to find appropriate key data for 300 mm silicon material, the simulator developed provides general capabilities for numerical, physically-based, high speed simulation of slip-free fast ramp processing steps.

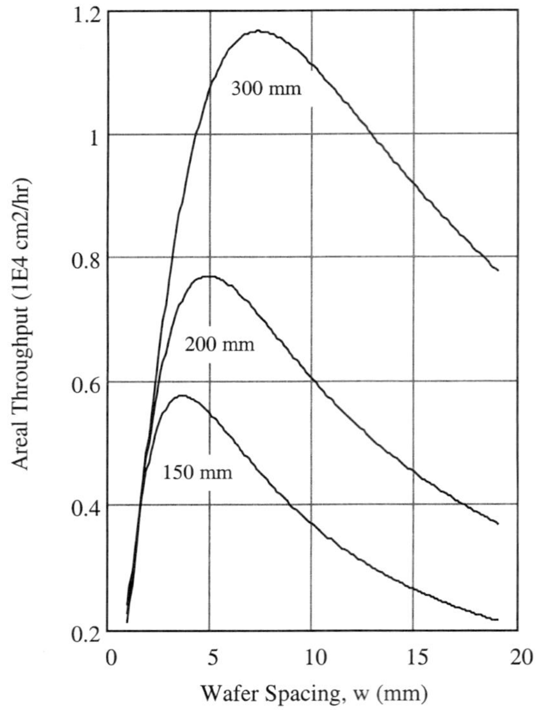

Fig. 10. Areal throughput per hour for a 2 h/1050°C high temperature process as a function of the face-to-face wafer spacing *w* for different wafer diameters and an effective furnace height of 500 mm.

References

[1] S.P. Timoshenko, J.N. Goodier, Theory of Elasticity, McGraw-Hill, New York, 1982.
[2] T. Fukuda, Jpn. J. Appl. Phys. 35 (1996) 3799.
[3] H.R. Huff, R.K. Goodall, Proc. 2nd Int. Symp. Adv. Sci. & Technol. of Si Mat., Nov. 25–29, 1996, Kona, Hawaii, USA, p. 322.
[4] S.K. Griffith, R.H. Nilson, R. Foote, 3rd Int. Workshop on 300 mm Diameter Wafers, SEMICON/West, July 10, 1995, SEMI, Mountain View, CA 94043, U.S.A.
[5] L.N.G. Filon, Trans. Roy. Soc. A 201 (1903) 67.
[6] K. Morezane, P.S. Gleim, J. Appl. Phys. 40 (1969) 4104.
[7] S.M. Hu, J. Appl. Phys. 40 (1969) 4413.
[8] S.M. Hu, S.P. Klepner, R.O. Schwenker, D.K. Seto, J. Appl. Phys. 47 (1976) 4098.
[9] P. Xin, Electrochem. Soc. Proc. 98–1 (1998) 660.
[10] I. Yonenaga, K. Sumino, Jpn. J. Appl. Phys. 21 (1982) 47.
[11] T. Fukuda, M. Koizuka, J. Appl. Phys. 74 (1993) 2420.
[12] H.-D. Chiou, Solid State Technol. 30 (1987) 77.
[13] M. Akatsuka, K. Sueoka, H. Katahama, N. Adachi, Electrochem. Soc. Proc. 98–1 (1998) 671.
[14] B. Leroy, C. Plougonven, J. Electrochem. Soc. 127 (1980) 961.
[15] M. Schrems, H.-J. Timme, P. Kücher, Private communication.

[16] R. Takeda, P. Xin, J. Yoshikawa, Y. Kirino, Y. Hosoki, N. Tsuchiya, O. Fujii, J. Electrochem. Soc. 144 (1997) L280.
[17] A. Fischer, Krist. Techn. 15 (1980) 1059.
[18] A. Fischer, Exp. Techn. Phys. 29 (1981) 95.
[19] S.M. Hu, J. Appl. Phys. 70 (1991) R53.
[20] H. Richter, G. Kissinger, A. Fischer, Proc. Symposium for the 20th Anniversary of the 145th Committee on Processing and Characterization of Crystals of JSPS, 1997, p. 34.

ELSEVIER Microelectronic Engineering 45 (1999) 225–236

MICROELECTRONIC ENGINEERING

Experimental verification of different slip generation models for 300 mm wafers processed in a fast ramp vertical furnace

G. Ritter[a,*], P. McHugh[a], G. Wilson[a], L. Funk[a], P. Zaumseil[b]

[a]Semitool, Inc., 655 West Res. Drive, Kalispell, MT 59901, USA
[b]Institute for Semiconductor Physics, W. Korsing Str. 2, Frankfurt (Oder), Germany

Abstract

This paper presents the results of 300 mm wafer slip tests conducted using Semitool's 300 mm Express furnace at process temperatures ranging from 900°C to 1150°C. X-ray diffraction topography is used to search for the presence of dislocations in the processed wafers. The experimental data is evaluated against the predictions of the wafer stress analysis code MacWafer/PCWafer [R. Nilson, S. Griffiths, MacWafer Computer Software, Sandia National Laboratories, Livermore, CA 94551-0969]. This code enables the application and evaluation of different slip generation models [M. Schrems et al., in: H.R. Huff, W. Bergholz, K. Sumino (Eds.), Semiconductor Silicon '94, The Electrochemical Society, N.J., 1994, pp. 1050–1063; B. Leroy, C. Plougonven, J. Electrochemical Society 127 (4) (1980) 961–970]. Instances of wafer slip are observed only at wafer support points. The occurrence of wafer slip is best described by the model attributable to reference [B. Leroy, C. Plougonven, J. Electrochemical Society 127 (4) (1980) 961–970 (1980)] assuming an oxygen precipitation level of about 4 ppm.

Keywords: Slip generation models; 300 mm Wafers; Fast ramp vertical furnace

1. Introduction

The conversion toward larger wafer size is mainly driven by economical reasons. The conversion to 300 mm diameter wafers will provide approximately 30% lower IC cost per die. However, there are a number of engineering challenges that face the silicon wafer and IC equipment suppliers in bringing 300 mm wafer utilization to a production level maturity. The control of plastic deformation in silicon wafers has been and will continue to be a major concern in the industry. When wafers are stressed at high temperatures, generation and gliding of dislocations within the crystalline material may occur, resulting in slip bands on the wafer surface.

The reasons for stress are temperature differences across the wafer, gravitational bending, contact stress at the wafer support points, and process induced stress (e.g., by thermal oxidation or stress

*Corresponding author. Tel.: +1-406-752-2107; fax: +1-406-752-5054.
E-mail address: gritter@semitool.com (G. Ritter)

0167-9317/99/$ – see front matter
PII: S0167-9317(99)00156-2

induced by layers of different materials deposited on the surface). Dislocations are generated and then moved through the wafer if the resulting shear stress along the preferred glide planes (the resolved shear stress) exceeds the critical yield shear stress (yield strength).

The amount of yield stress is still the most uncertain factor in all efforts trying to model and to control the plastic deformation in Si wafers. There are different experimentally obtained models concerning dependence of yield strength on different parameters, e.g., temperature, oxygen concentration and oxygen precipitation. They provide extremely different, sometimes even contradictory, predictions concerning slip generation in wafers during thermal processing. Therefore, experimental tests are required to determine the best yield stress criterion to use for a specific process and type of material.

For this purpose, we used the MacWafer Computer Code developed at Sandia National Laboratories [1]. MacWafer considers both thermomechanical stress and gravitational bending stress at the contact points. The predictions of two different slip criteria [2,3] were compared with experimental XRD analysis results. In our case, the occurrence of wafer slip is best described by the model attributed to [3] assuming an oxygen precipitation level of about 4 ppm.

2. Mechanical stress and defect generation in vertical batch processing

2.1. Thermal stress

Thermal stress in silicon wafer batches during furnace treatment was already studied by Hu as early as 1969 [4]. The radial temperature gradient across the wafer in the rotationally symmetric hot wall system can be modeled by a power function

$$(T(r) - T_c)/(\Delta T) = (r/R)^n \quad (1)$$

$$\Delta T = T_e - T_c \quad (2)$$

where $T(r)$ is the temperature at the radial position r. Parameter R is the radius of the wafer, T_c and T_e the temperatures at center and wafer edge, respectively. The silicon wafer is usually represented as an isotropic thin, circular disk with uniform temperature through the wafer thickness in the z-direction. The stress tensor then simply contains only two different components, the radial stress σ_r, and the tangential stress σ_θ. For a temperature profile as described in (1), those components can be expressed by the following formulas

$$\sigma_r = \alpha E\, \Delta T\{1 - (r/R)^n\}/(n + 2) \quad (3)$$

$$\sigma_\theta = \alpha E\, \Delta T\{1 - (n + 1)(r/R)^n\}/(n + 2) \quad (4)$$

where alpha (α) is the coefficient of thermal expansion, and E is Young's modulus. For furnace processes, it has been shown that the temperature profile is best described by a parabolic function ($n = 2$).

The maximum shear stress $\tau_{s,max}$ in the wafer plane resulting at the wafer edge is expressed as

$$\tau_{s,max} = 1/2[\sigma_r - \sigma_\theta] \quad (5)$$

$$\tau_{s,max} = \alpha E\ \Delta T/4 \quad (6)$$

It is interesting to note that the maximum shear stress (at the wafer edge) does not depend on wafer radius, but increases linearly with increased temperature difference and depends on Young's modulus and the thermal expansion coefficient. For silicon wafers in this parabolic temperature profile, the thermal stress caused by every 1^0C temperature difference is equal to approximately 0.13 MPa (following reference [6] $\alpha = 4 \times 10^{-6}$ 1/K, $E = 131$ GPa).

Temperature differences across the wafer during heating up or cooling down in a hot wall reactor are modeled in MacWafer [1] considering both heat radiation and heat conduction [5],

$$\Delta T = \Gamma(\delta T/\delta t) \quad (7)$$

with the factor Γ given by

$$\Gamma = \frac{\rho c d^2}{16\{k + 4.6\sigma T^3 w(w/t)(w/d)^{-1/4}\}} \quad (7a)$$

where $\delta T/\delta t$ = ramp rate, d = wafer diameter, w = wafer spacing, t = wafer thickness, k = thermal conductivity of Si, σ = Stefan-Boltzmann constant, c = specific heat of Si, ρ = density of Si, and T = average wafer temperature. Expression (7) allows one to relate wafer and batch geometry (w/t, w/d) and furnace parameters (T, $\delta T/\delta t$) with temperature difference across the wafer (ΔT). Combining (6) and (7) allows one to estimate the maximum thermal shear stress across the wafer during heating or cooling in the furnace.

2.2. Gravitational bending stress

Large wafers held in a horizontal position in a vertical furnace encounter bending displacement due to the force of their own weight and so bending stresses must also be considered. The bending shear stress τ_g induced by gravitation is generally proportional to the square of wafer radius R, and inversely proportional to the thickness t [6]

$$\tau_g = G\rho g R^2/t \quad (8)$$

where g = gravitational acceleration constant, ρ = density of silicon, G = geometry factor depending on the support geometry. Note, that σ_g does not depend on Young's modulus E.

Gravitational bending stress has been calculated for different support geometries including ring support at different radii, and symmetrical and asymmetrical 4 and 3 point supports at different support radii [6,7]. Maximum bending shear stress occurs at the wafer center for ring supports and is minimum if the support ring (or support points) are located at approximately 70% of the wafer radius. In the case of 300 mm wafers, the maximum bending stress for an outer ring support is 0.41 MPa (in the wafer center), while for a 70% inner ring radius the stress is reduced to 0.11 MPa. For comparison, a symmetrical four point support at the wafer edge results in a stress of 1.59 MPa, a symmetrical three point support at the wafer edge results in a stress of 2.32 MPa, an asymmetrical three point support at the edge has a 3.86 MPa bending, and a symmetrical three point support at 70% inner radius has about 1.04 MPa of bending stress. Therefore, for these support options, the bending stress of 300 mm wafers varies between a minimum of 0.11 MPa to a maximum of 3.86 MPa, respectively. The amount of this shear stress component in 300 mm wafers corresponds to an equivalent stress due to temperature differences between 1°C to 27°C, respectively.

2.3. Slip free criteria

Plastic deformation due to mechanical stress (gliding of existing dislocations) starts if the resolved shear stress exceeds the critical yield shear stress (yield strength) τ_{yield} of the material. Therefore, a criterion for avoiding generation of slip lines in the wafer can be formulated as

$$(\tau_{s,max} + \tau_{g,max}) < \tau_{yield} \tag{9}$$

Using formula (6), the criterion for slip avoidance can be expressed in terms of allowable temperature difference ΔT_{allow} across the wafer

$$\Delta T_{allow} < 4(\tau_{yield} - \tau_{g,max})/(\alpha E) \tag{10}$$

or using formula (7) and (7a), in terms of allowable ramp rates $(\delta T/\delta t)_{allow}$

$$(\delta T/\delta t)_{allow} < 4(\tau_{yield} - \tau_{g,max})/(\alpha \Gamma E) \tag{11}$$

Generally, the critical yield shear stress decreases exponentially with increasing temperature, decreases with increasing concentration of oxygen precipitates, and increases with increasing amount of interstitial oxygen. However, in addition to these parameters, many other factors (e.g., wafer shape, defect density due to wafer history, etc.) affect τ_{yield}. There are many different experimentally obtained models of τ_{yield}. In this paper, only two of the most popular [2,3] will be compared.

The highest concern regarding the different slip free criteria used here is the uncertainty in τ_{yield}. It is very important to note, that all models for plastic deformation of silicon wafers due to multiplication and gliding of dislocations require already existing 'generator' dislocations. The energy required for generation and growing of new dislocations (by climbing) is significantly higher than the energy demanded for multiplication and gliding. Unfortunately, no independently obtained confident data are known. Nevertheless, there are many different models for dislocation generation during thermal processing. Generally, initial dislocations can be created by heterogeneous or homogeneous nucleation [8]. At commonly used yield conditions, plastic deformation by creation of gliding lines occurs only if enough dislocations have already been created in the region of interaction with the applied mechanical stress field.

Heterogeneous nucleation of dislocations occurs when defects from external sources interact with the applied mechanical stress field., e.g., grown-in during crystal growth, or punched out by growing precipitates, end-of-range defects introduced by ion implantation, mechanical scratches (very often found at the wafer edge), and so on. One possible source of dislocation may be the very high contact stress at the support point of wafers [6] which can reach several hundred MPa.

Homogeneous nucleation occurs by direct condensation of self-interstitials. This type of nucleation requires a much higher stress level than the glide force exerted by the externally applied stress field and must exceed the counteracting line tension force that is inversely proportional to the radius of curvature of the dislocation [8–10]. Examples of sources for self interstitials allowing the process of homogeneous nucleation to occur are self interstitial rich outer regions in 300 mm crystals, oxygen precipitates, and injection of point defects during film growth.

The concentration of other impurities, (e.g., interstitial oxygen or very high concentration of Boron atoms) can block already existing dislocations (pinning effect). It has been shown recently [11], that the equivalent dislocation pinning stress due to oxygen concentration of 5×0^{17} Atcm^{-3} (10 ppma) reaches values between 0.8 MPa at 700°C, and 0.4 MPa at 1200°C. For oxygen concentrations of 1.5×10^{18} Atcm^{-3} (30 ppma), the corresponding pinning stresses are even higher at 2.4 MPa and 1.1 MPa, respectively. These values are even higher than gravitational bending stresses and yield stress at very high temperatures.

It is interesting to note that the critical yield values in the model of [3] differs at high temperatures from the yield values in the model of [2] by exactly those amounts. Assuming that the 300 mm material used in [2] had a much higher initial oxygen concentration than the material used in [3], one could speculatively explain the much higher yield strength at higher temperature observed in [2] due to higher pinning stress in this material. On the other hand, this extremely oxygen rich material could create higher concentrations of oxygen precipitates at lower temperatures than in [3], and therefore increase the intrinsic stress in the material lowering the yield stress in comparison to [3].

It has been reported very recently [11,12] that the yield stress in highly Boron doped 300 mm silicon wafers ($\rho=0.9\ldots10$ mΩcm) is significantly higher than in low doped material ($\rho=10\ldots15$ Ωcm), even with extremely high oxygen concentrations of about 1.6×10^{18} Atcm^{-3}.

3. Experimental procedure

This paper presents the results of wafer slip tests conducted using Semitool's 300 mm Express Furnace. The design of the furnace is schematically presented in Fig. 1. Two different wafer support configurations are investigated in processes at temperatures ranging from 900°C to 1150°C.

X-ray diffraction topography (XRD) is used to search for the presence of dislocations in wafers processed. Fourier Transform Infrared analysis (FTIR) has been employed for measurement of interstitial Oxygen concentration and Infrared Scattering Tomography (IST) has been used to characterize wafers regarding precipitates.

In the experiments, 300 mm wafers with an initial Oxygen concentration in the region between 25 ppma and 31 ppma and a low Boron doping level (10–15 Ωcm) have been used. The most important test parameters are reported in Table 1.

Tests at temperatures below 950°C employ a conventional tower design, in which wafers are supported at three edge positions separated by 93.3°. With this tower design, wafer pitch values of 22 mm, 11 mm, and 5.5 mm were studied corresponding to the use of 30, 60, and 121 wafers in the tower, respectively. The corresponding normalized wafer spacing (ratio of wafer distance to wafer diameter, w/d) was therefore 0.073, 0.037, and 0.018, respectively. In the tests O and L, after a stabilization of the temperature in the wafer batch at 400°C, the temperature was increased linearly at 30°C/min to 900°C and after a stabilization time ramped down at 20°C/min. In an other test (I), the wafers were very slowly heated up and cooled down: specifically from 400°C to 800°C at 10°C/min, from 800°C to 900°C at 5°C/min and finally to 950°C at 1.6°C/min. Cooling down was performed similarly.

Thermal process tests at temperatures above 950°C use a test tower design in which wafers are supported at three interior points located at a radial position equal 70% of wafer radius. These points have an angular separation of 120°. In the first test (A), only one wafer has been loaded in order to test

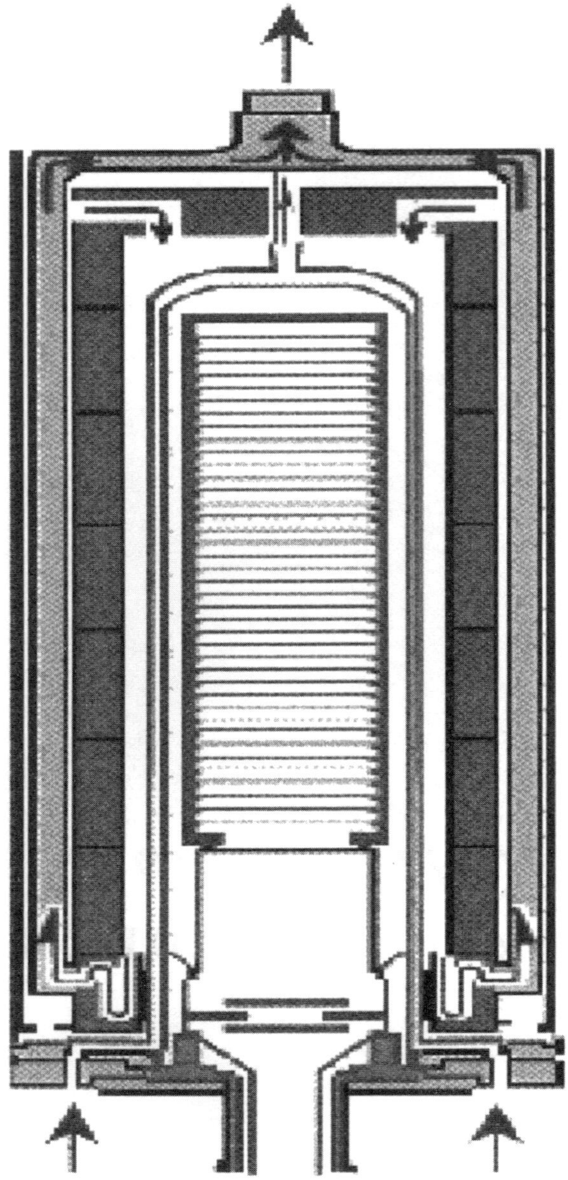

Fig. 1. Vertical furnace for 300 mm wafers (EXPRESS 300, Semitool, Inc.).

the effects of gravitational bending stresses while minimising temperature difference across the wafer (normalized wafer spacing $\gg 1$). In the remaining tests (C, D, E) the wafer pitch value was 11 mm ($w/d = 0.037$). The heating cycle again started at 400°C. The temperature was then increased with different ramping rates, and after stabilizing at the prescribed final temperature (1050°C, 1150°C), cooled down. Different combinations of critical/uncritical heating up with critical/uncritical cool

Table 1
Main parameters used in slip tests

Test/parameter	A	C	D	D1	E	I	L	O
Max Temp. [°C]	1150	1150	1050	1050	1050	950	900	900
Heat $\delta T/\delta t$ [°C/min]	12.7/1.6	2.5/1.6	12.7	30	12.7	10/5/1.6	30	30
Cool $\delta T/\delta t$ [°C/min]	1.6/7	1.6/2.5	2.5	2.5	20	1.6/5	20	20
w/d	$\gg 1$	0.037	0.037	0.037	0.037	0.073	0.037	0.018
$\tau_{g,max}$ [Mpa]	1.04	1.04	1.04	1.04	1.04	3.86	3.86	3.86

down were used. The main information from these tests are shown in Table 1 along with the previously discussed tests.

4. Predictions of different slip criteria

The software MacWafer [1] has been employed which uses different models for critical shear stress (e.g., Schrems et al. [2] commonly known as the TOSHIBA/TEL criterion, and Leroy/Plougonven (indicated as LP) [3] for oxygen precipitation concentrations of 0 ppma, 2 ppma, 4 ppma, and 8 ppma, respectively) to predict allowable ramping rates in the region between 500°C and 1200°C for 300 mm wafers. The corresponding bending stresses due to the different support geometries (compare with Table 1) were considered.

At 900°C, the critical yield stress estimated by the model of [2] is about 4.5 MPa while the corresponding values estimated by the model of [3] vary between 8 and 4 MPa for oxygen precipitations equal to 2 ppma and 8 ppma, respectively. In this case, the high value of bending stress (3.86 MPa for the asymmetric three point support at wafer edge) is critical, but still tolerable in all cases.

At 1150°C, the critical stress values using the model of [2] are reduced to 2.5 MPa and using the model of [3] are even smaller reaching 1.5 MPa and 0.4 MPa for oxygen precipitations equal to 2 ppma and 8 ppma, respectively. In any case, the three point support at the wafer edge generates sufficient bending stress to cause gliding even without any temperature difference. Therefore, the alternative three-point support located at 70% wafer radius with a reduced bending stress of about 1.04 MPa, was used in these cases. At these conditions, the slip free criterion of [2] would be fulfilled, and no slip generation is predicted. However, the slip criterion of [3] would predict no slips only for very low concentrations of oxygen precipitates.

5. Results

The slip free criteria has been estimated in terms of allowable ramp rates in dependence on the processing temperature for the individual test conditions. The actual ramp rates were measured by profile thermocouples and then compared with the maximum allowable rates predicted by the different criteria. Examples of ramp rate curves for some tests are demonstrated in Figs. 2–5. The predictions regarding slip creation during heating and cooling in the individual tests are summarized in Table 2.

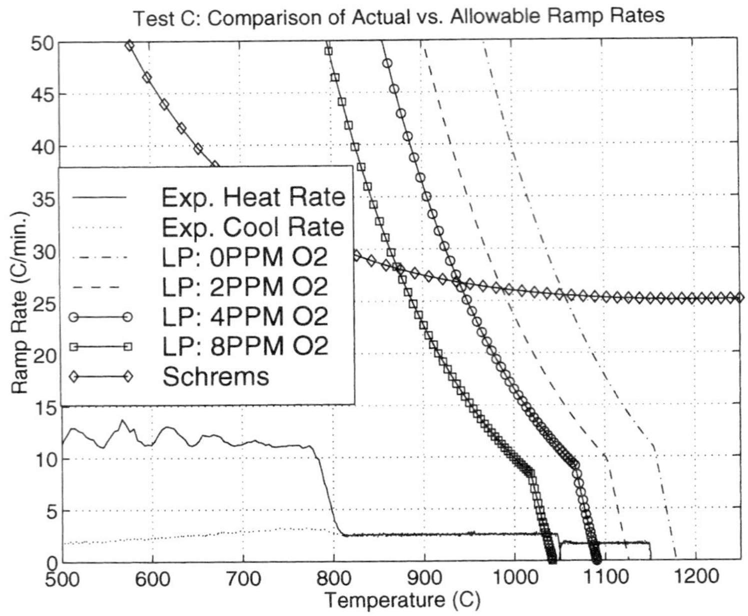

Fig. 2. Allowable ramp rates in comparison with measured rates for test C.

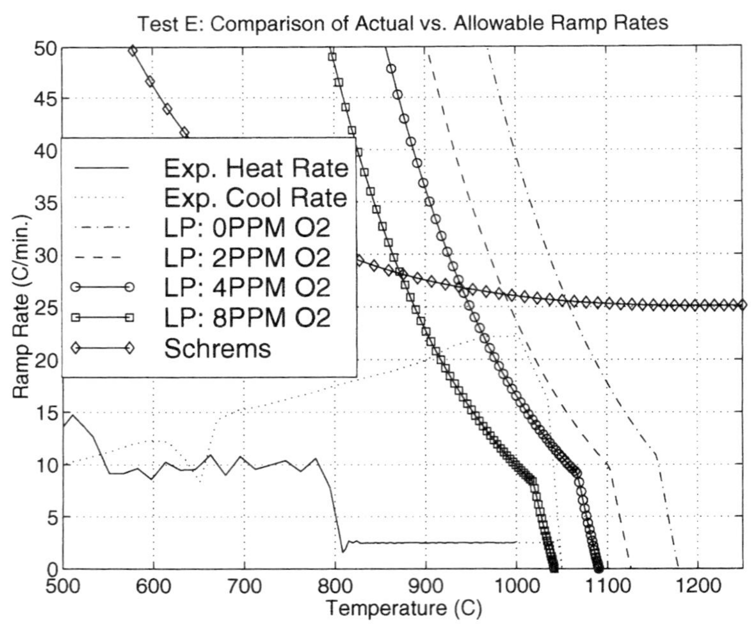

Fig. 3. Allowable ramp rates in comparison with measured rates for test E.

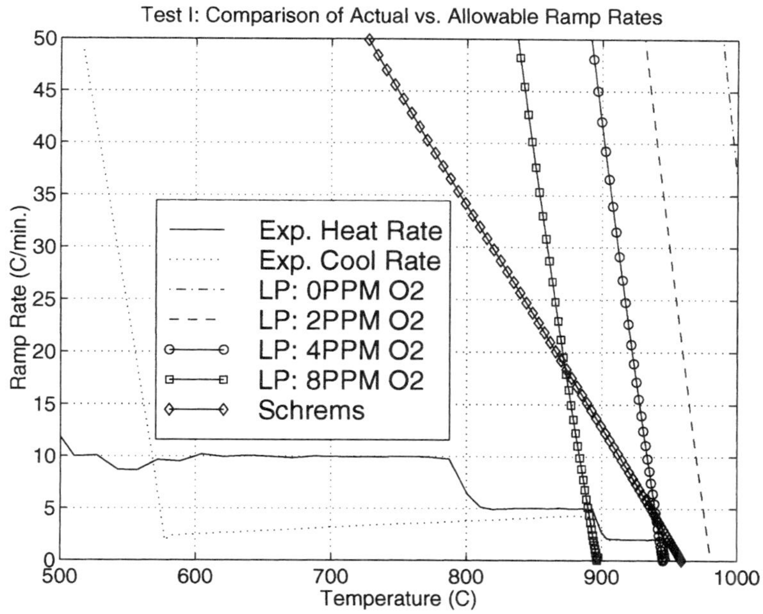

Fig. 4. Allowable ramp rates in comparison with measured rates for test I.

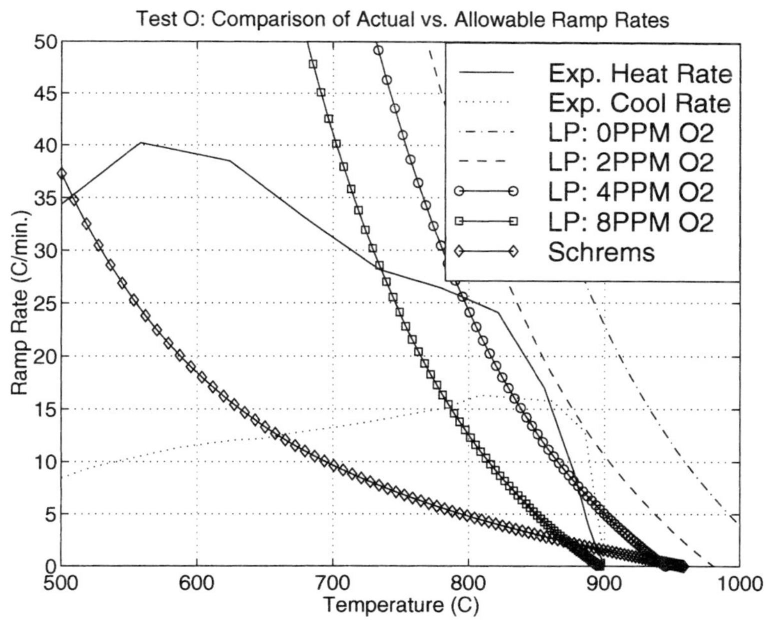

Fig. 5. Allowable ramp rates in comparison with measured rates for test O.

Table 2
Slip criteria of test samples described in Table 1 compared with XRD results[a]

Test	A	C	D	D1	E	I	L	O
Schre.	N/N	N/N	N/N	Y/N	N/N	N/N	Y/Y	Y/Y
LP-0	N/N	N/N	N/N	N/N	N/N	N/N	N/N	N/N
LP-2	Y/Y	Y/Y	N/N	N/N	N/N	N/N	N/N	N/N
LP-4	Y/Y	Y/Y	Y/N	Y/N	N/Y	Y/Y	N/N	Y/Y
LP-8	Y/Y	Y/Y	Y/Y	Y/N	Y/Y	Y/Y	Y/Y	Y/Y
XRD	Yes	Yes	Yes	Yes	Yes	No	No	Yes
Result	(15 mm)	(15 mm)	(2 mm)	(2 mm)	(2 mm)			(0.5 mm)

[a] Row indicated Schre. shows predicted slip generation according with [2] at heating/cooling. LP-0 to LP-8 are predicted slip generation according to [3] for Oxygen precipitations of 0 ppma, 2 ppma, 4 ppma, and 8 ppma, respectively, during heating/cooling. N = no slips, Y = slips predicted.

After the annealing experiments, the wafers tested were cut in smaller parts in accordance with Fig. 6. Parts 2, 3, and 4 contain the critical support points, at which first dislocations are expected. X-ray diffraction (Lang-transmission analysis) was used to check dislocations and slip lines. It was evident in all cases studied in this experiment, that instances of slips occurred only at the most critical support points. The slip line lengths varied between 15 mm in the most critical cases to only 0.5 mm in the less critical tests. Examples of XRD analysis of the most critical samples are demonstrated in Figs. 7 and 8.

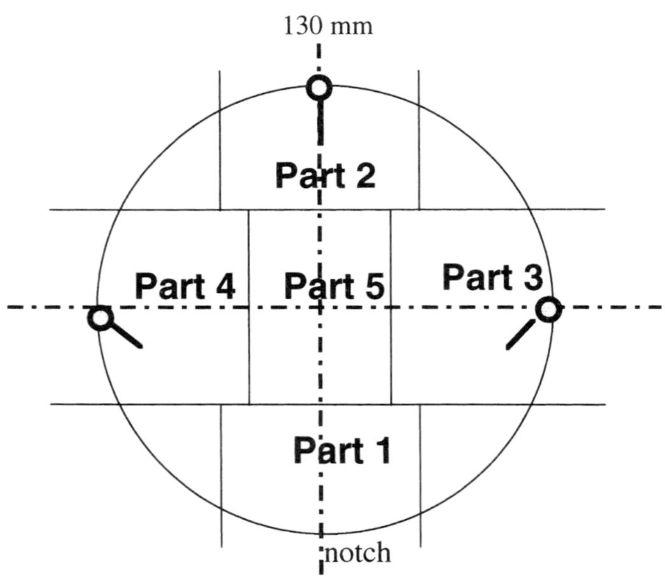

Fig. 6. Cutting of samples for XRD dislocation analysis.

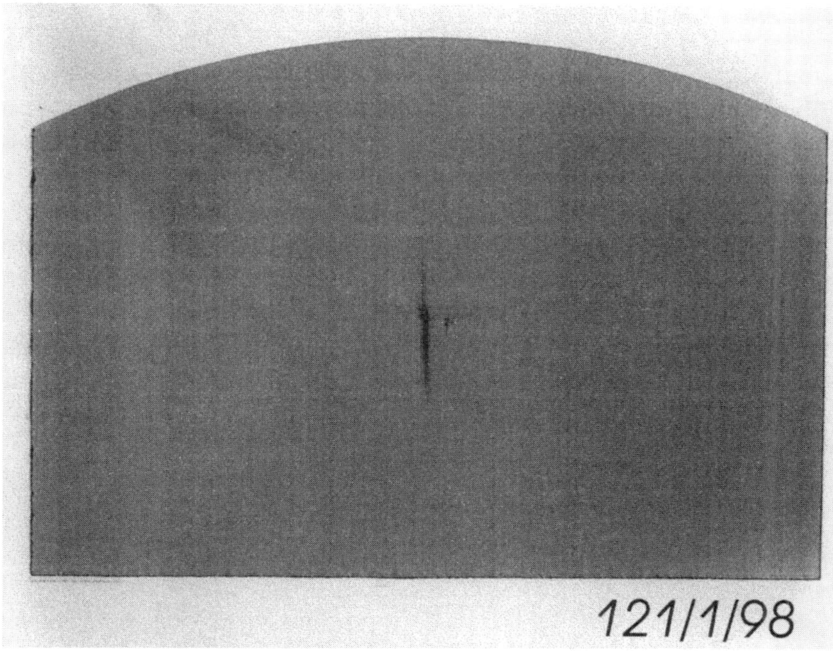

Fig. 7. XRD dislocation analysis of test sample A, part 2.

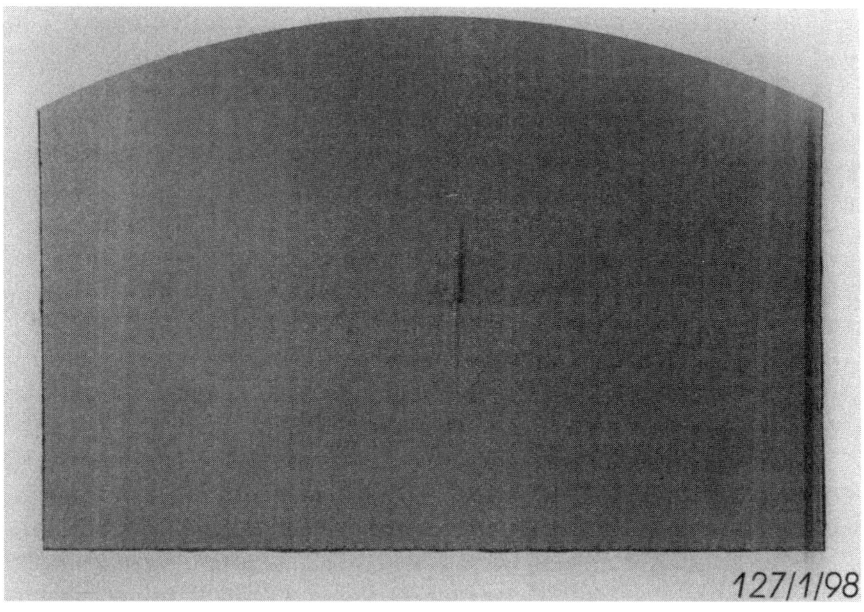

Fig. 8. XRD dislocation analysis of test sample C, part 2.

6. Discussion and conclusions

Comparing the XRD results concerning creation of slip lines of individual test samples with the predictions of different slip criteria [2,3], shown in Table 2, it can be concluded that the occurrence of wafer slip in 300 mm wafers are best predicted by the model of [3] assuming a mean precipitation level of about 4 ppma. The latter value is consistent with the results of IR analysis, as well.

It has been shown that according to theoretical considerations, gravitational stress is significant for 300 mm wafers at higher temperatures and therefore, optimization of support geometry in vertical batch reactors is an essential issue. Theoretically, wafer support with minimum bending stress (e.g. inner rings with 0.7 wafer radius) of only 0.11 MPa should be able to allow slip free processing of wafers even at temperatures much higher than 1100°C. However, according to recently published results [6], ring supports and wafer holders containing more than three support points are difficult to realize because of deformation of both the holder and the wafer. In practice, such wafer holders are reduced with high probability to only three supporting points.

In order to be sure to avoid creation of slips in 300 mm wafers at temperatures higher than 1100°C, wafers with very high amounts of impurities to pin the motion of dislocations (e.g., high concentration of interstitial oxygen or very high concentration of Boron) should be used.

References

[1] R. Nilson, S. Grifffiths, MacWafer Computer Software, Sandia National Laboratories, Livermore, CA 94551-0969.
[2] M. Schrems et al., in: H.R. Huff, W. Bergholz, K. Sumino (Eds.), Semiconductor Silicon'94, The Electrochemical Society, N.J, 1994, pp. 1050–1063.
[3] B. Leroy, C. Plougonven, J. Electrochem. Soc. 127 (4) (1980) 961–970.
[4] S.M. Hu, J. Appl. Phys. 40 (1969) 4413.
[5] R.H. Nilson, S.K. Griffiths, Scaling Batch Processes for Large Diameter Wafer, in: H.R. Huff, K. Torres (Eds.), Second SEMATECH Large Diameter Wafer Thermal Issues Conference, Sept., 25 1996.
[6] P. Xin, in: H.R. Huff, U. Goesele, H. Tsuya (Eds.), Proc. Semiconductor Silicon, vol. 98-1, The Electrochemical Society, 1998, p. 660.
[7] T. Fukuda, Jpn. J. Appl. Phys. 35 (1996) 3799.
[8] J. Vanhellemont, in: H.R. Huff, U. Goesele, H. Tsuya (Eds.), Proc Semiconductor Silicon, vol. 98-1, The Electrochemical Society, 1998, p. 997.
[9] J. Vanhellemont et al., J. Appl. Phys. 62 (1986) 2170.
[10] J. Vanhellemont et al., J. Appl. Phys. 62 (1986) 2176.
[11] M. Akatsuka et al., in: H.R. Huff, U. Goesele, H. Tsuya (Eds.), Semiconductor Silicon, vol. 98-1, The Electrochemical Society, (1998), p. 671.
[12] K. Hikazutani et al. (Fujitsu Ltd., Japan), 193rd Meeting of The Electrochemical Society, San Diego, May 3–8, 1998, Abstr. No. 518, 1998.

ELSEVIER

Microelectronic Engineering 45 (1999) 237–246

MICROELECTRONIC ENGINEERING

Challenges and current status in 300 mm rapid thermal processing

M. Glück*, W. Lerch, D. Löffelmacher, M. Hauf, U. Kreiser

STEAG RTP Systems GmbH, Daimlerstrasse 10, D-89160 Dornstadt, Germany

Abstract

Rapid Thermal Processing (RTP) will be indispensable to meet the process requirements of future device and integrated circuit (IC) generations. The paper reviews several approaches to manage the challenges of single-wafer 300 mm heat treatment, including the most important issues in RTP system technology: temperature measurement and accuracy, fast ramp capability, process repeatability, ultra-shallow junction formation and process uniformity for various applications in concurrent IC manufacturing. The paper also focuses on the substantial improvement of key system features. The implementation of a new lamp intensity modulation-based and emissivity independent temperature measurement technique yields a measurement uncertainty of ±1.5°C within an emissivity range from 0.2 to 0.8 for key process applications.

Keywords: Rapid Thermal Processing; RTP system technology

1. Introduction

The world-wide market for consumer electronics is predicted to increase to a volume above 200 billion US$ by the year 2000 with an expected share of more than 95% for silicon devices and circuits [1,2]. Most of these are CMOS microcontrollers and DRAMs which are monolithically integrated in outstanding numbers. With the continuous down-scaling of all relevant device parameters in conjunction with the inevitably increasing wafer size and the more and more stringent process requirements, typical furnace annealing steps will completely be replaced by single-wafer processes [3]. RTP has become a key fabrication technology for advanced integrated circuits. The growing field of applications includes Ti and Co silicide formation, implant annealing and dopant activation (RTA), shallow junction formation and reflow of deposited oxides [4]. The reduced thermal budget, improved ambient control, fast cycle times and the compatibility with single-wafer processing will open a continuously broadening field of new applications, e.g. the formation of ultra-thin gate dielectrics (RTO, RTN) and rapid thermal chemical vapor deposition (RTCVD).

*Corresponding author. Tel.: +49-7348-981-287; fax: +49-7348-981-249.
E-mail address: m.glueck@steag-ast.de (M. Glück)

0167-9317/99/$ – see front matter
PII: S0167-9317(99)00157-4

2. System performance

The experiments are performed in STEAG RTP Systems' latest high-end 200 and 300 mm RTP production tool AST 3000. The system features double-side tungsten halogen lamp heating (56 linear lamps providing a maximum total heating power of 170 kW) including independent lamp control of both, top and bottom heating lamps, to enable high ramp-up rates (up to 350 K/s for 200 mm and 250 K/s for 300 mm wafers) as well as to account for possible pattern-related temperature non-uniformities across typical production-type wafers, see Fig. 1. The wafer is positioned on a rotating quartz wafer tray with 6 pins which prevent excess heat transfer losses causing undesirable temperature non-uniformities. The process gases enter the leak-tight reactor chamber at the backside. Gas distributor plates create a homogeneous laminar gas flow above the inserted wafer. Wafer rotation is obtained by generating a process gas cushion below a centered, but moveable quartz plate. Nozzles arranged outside the quartz plate allow a software controlled acceleration and slowing-down of the rotating quartz tray via additional gas injection (rotation speed up to 100 rpm enabled). A detailed system description is given elsewhere [5]. The impact of rotation will be discussed in the following.

The stringent process uniformity and repeatability requirements demand a continuous improvement in wafer temperature control. The former patent pending temperature measurement technique evaluating emitted radiation at 2.7 μm wave-length which was transferred across special filter glasses having a very slow Si–OH absorptivity was substantially modified. The wafer itself is the most important and unpredictable variable as its optical constants dominate its thermal response. Since the wafer is not in thermal equilibrium with the heat sources at almost any part of the RTP cycle, it is the wafer temperature which has to be controlled as precisely as possible. By the use of active intensity-modulated lamp light an emissivity independent, compared to former attempts in ripple

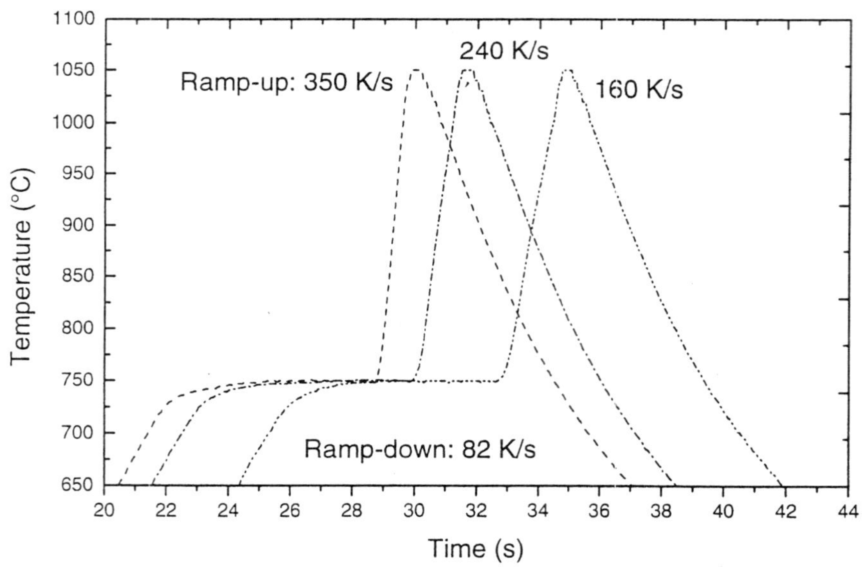

Fig. 1. Pyrometer read-out for three different ramp up rates (160 K/s, 240 K/s, 350 K/s) with extended fast-cooling of − 82 K/s.

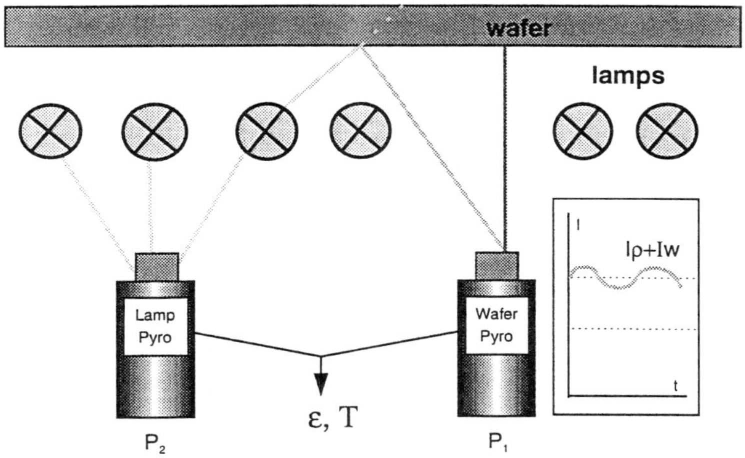

Fig. 2. Measurement principle of the light intensity modulated temperature measurement technique. By simultaneously monitoring the modulated lamp light (pyrometer P_2) and the emitted wafer radiation (P_1) the superposition of wafer radiation and reflected lamp light can be separated. As a consequence wafer emissivity ϵ and temperature T can be determined.

pyrometry [6,7] substantially improved new temperature measurement technique for double-side heated RTP systems has recently been implemented. The improvements depend on lots of details in optics, electronics, signal processing and appropriate modeling, being described by Hauf et al. in another publication [8]. The measurement principle is sketched in Fig. 2. The wafer temperature is calculated from the output signals of the two pyrometers, one focussed on the lamps, the other one focussed on the wafer backside. Using a sophisticated model which takes into account the dominating radiant contributions and interactions in the chamber enables a robust and universal interpretation of the measured signals for all wafers within an emissivity range from 0.2 to 0.8 providing a measurement uncertainty of ±2°C for various backside reflectivities (see Fig. 3). Extensive calibration procedures using TC coupled wafers will no longer be necessary as the system manages to account for changes in emissivity and reflectivity of different wafer types being strongly influenced by backside coating, wafer roughness, substrate doping and process temperature thus enabling substantially reduced system up-times.

3. Processes

A set of evaluation experiments was performed to investigate temperature uniformities during steady-state and ramp-up under realistic conditions for 200 and 300 mm operation. It is obvious that process uniformity depends on respective temperature and flow distributions in the chamber. The dominating influence however is radiation: unlike emission and absorption properties of the wafers and the highly-reflective chamber walls yield non-uniform primary radiation flux densities [9,10]. The inherent process uniformity was analyzed by implant annealing of As implanted 200 and 300 mm wafers ($1 \cdot 10^{16}$ cm^{-2}, 20 keV, RTA: 1000°C, 30 s). On 200 mm wafers, a temperature variation of $\Delta T = \pm 2.4$°C and a standard deviation of $\sigma = 1.1\%$ was observed without any optimization of the

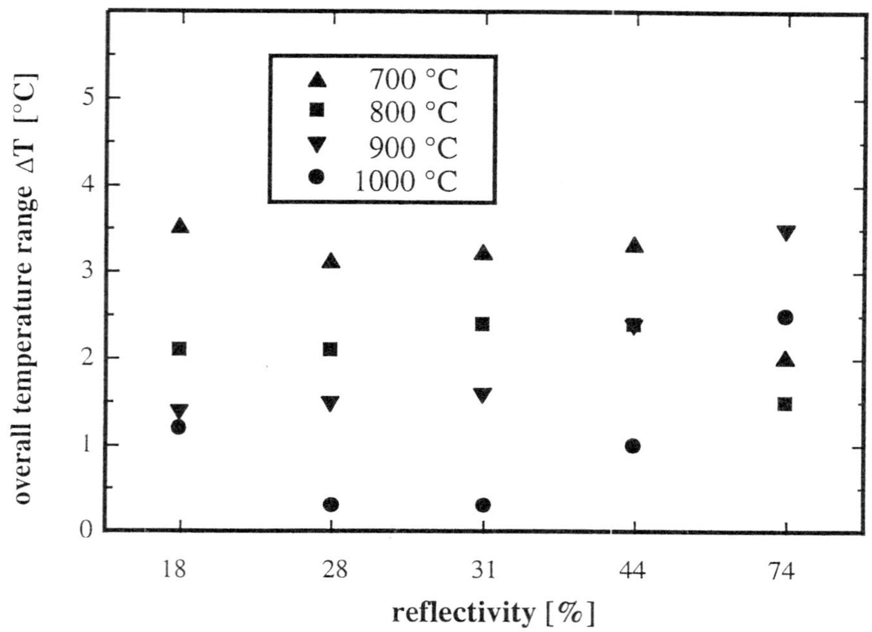

Fig. 3. Maximum overall temperature range ($T_{max} - T_{min}$) of multiple TC measurements on 200 mm wafers (edge exclusion 3 mm) as a function of annealing temperature for different wafer backsides with varying reflectivity.

lamp correction tables (LCT), indicating a very homogeneous radiation distribution and basic uniformity of the process chamber, an important prerequisite for off-line OPUS-assisted optimization [11,12]. An optimized LCT results in $\Delta T = \pm 1.2$°C and $\sigma = 0.77\%$.

For 300 mm wafers, the inherent chamber uniformity is worse due to the reduced chamber-to-wafer size ratio, yielding $\Delta T = \pm 10.4$°C and $\sigma = 4.3\%$ without lamp correction and substrate rotation. Using wafer rotation for the compensation of chamber non-uniformities yields a reduction to $\Delta T = \pm 5.0$°C and $\sigma = 3.6\%$, still without applying any lamp correction. A rotation speed of 20 rpm is sufficient. Further OPUS optimization finally yields an excellent $\Delta T = \pm 1.3$°C and $\sigma = 0.76\%$. For all measurements, an edge exclusion of 5.5 mm was applied according to the specification of the implantation process. A profound reliability testing showing an excellent long-term stability occurred within the I300I programme. A total amount of 14 339 wafers was processed within a 21 day marathon test in a production-like environment. The repeatability of the sheet resistance values after RTA (As implant: 25 keV, $1 \cdot 10^{16}$ cm^{-2}; RTA: 1000°C, 30 s) correlates with a total temperature variation of ±0.25°C. During the entire period no failure or production interrupt occurred. A brief summary of further representative process results for concurrent key applications is given in Table 1.

4. Fast ramp performance for ultra-shallow junction formation

One of the main technological challenges is the formation of ultra-shallow junctions with low parasitic series resistance for self-aligned silicides and minimized short-channel effects. According to the 1997 National Technology Roadmap for Semiconductors (NTRS) the targeted junction depths will

Table 1
Summary of representative process results for concurrent RTP key applications

Process	Wafer size [mm]	Process parameters	Rotation	Analysis	Results
RTA	200 mm	As, 25 keV $1 \cdot 10^{16}$ cm^{-2} 1100°C, 10 s, N_2/O_2	20 rpm	4-point probe	$R_{sh} = 36.6\ \Omega/sq.$ $\sigma = 0.46\%$ $\Delta R_{sh} = 0.98\ \Omega/sq.$ $\Delta T = \pm 0.75°C$
RTA	300 mm	As, 20 keV $1 \cdot 10^{16}$ cm^{-2} 1000°C, 30 s, N_2/O_2	20 rpm	4-point probe	$R_{sh} = 75.95\ \Omega/sq.$ $\sigma = 0.67\%$ $\Delta R_{sh} = 2.26\ \Omega/sq.$ $\Delta T = \pm 1.2°C$
RTO	200 mm	bare Si 1075°C, 15 s, O_2	20 rpm	ellipsometry	$t_{ox} = 58.7$ Å $\sigma = 0.30\%$ $\Delta t_{ox} = 1$ Å $\Delta T = \pm 1.5°C$
RTO	300 mm	bare Si 1150°C, 30 s, O_2	no	ellipsometry	$t_{ox} = 134.3$ Å $\sigma = 1.06\%$ $\Delta t_{ox} = 2.1$ Å $\Delta T = \pm 4.3°C$
SPA	200 mm	As, 20 keV $1 \cdot 10^{16}$ cm^{-2} 1050°C, flash, RU: 200 K/s, N_2/O_2	no	4-point probe	$R_{sh} = 129.4\ \Omega/sq.$ $\sigma = 0.53\%$ $\Delta R_{sh} = 4.86\ \Omega/sq.$ $\Delta T = \pm 2.0°C$
C49 → C54 $TiSi_2$	200 mm	20 nm Ti 720°C, 30 s, N_2	20 rpm	4-point probe	$R_{sh} = 2.3\ \Omega/sq.$ $\sigma = 3.1\%$

be in the order of a few tens of nanometers for the up-coming IC generations [13]. In 1997, Shishguchi et al. [14] already reported that ramp rates up to 400 K/s can significantly decrease the junction depth but the reproducibility in the ramp-up results was limited due to high levels of O_2 background concentrations. Earlier publications [15–17] have already shown that the annealing ambient plays a crucial role during ultra-shallow junction formation. This paragraph summarizes the advantages of high ramp-up rates and fast cool-down rates to find an optimum rapid thermal annealing process for ultra-shallow junctions based on $^{11}B^+$ (1 keV, $1 \cdot 10^{15}$ cm^{-2}) and $^{49}BF_2^+$ (2.2 keV, $1 \cdot 10^{15}$ cm^{-2}) [19]. The wafers (⟨100⟩, n-type $\rho \approx 10$–20 Ωcm) were processed on a Varian VIISion-80 Plus ion implanter. The detailed experimental conditions and additional results have been described in [18]. For all processes, the thermal stabilization step at 750°C for 10 s, which is commonly used to improve uniformity for lightly doped wafers, is eliminated yielding only a slight impact on junction depth and sheet resistance. For all experiments the remaining oxygen concentrations in the N_2 ambient were 60 to 65 ppm.

Downey et al. show that an increase by a factor of 4 in the ramp-up rate from 40 K/s to 160 K/s decreases the junction depth of a 1 keV B implant ($1 \cdot 10^{15} cm^{-2}$) from 940 Å to 880 Å (6%, determined at a dopant concentration of 10^{17} cm^{-3}) [19]. The shorter the soak times, the more pronounced is the impact of the ramp-up and cooling rates. The junction depth for a ramp-up rate of 240 K/s is 910 Å probably due to a SIMS measurement error or processing error, as additional results demonstrate. The impact of an increased ramp-down rate (86 K/s) on the junction depth is shown in

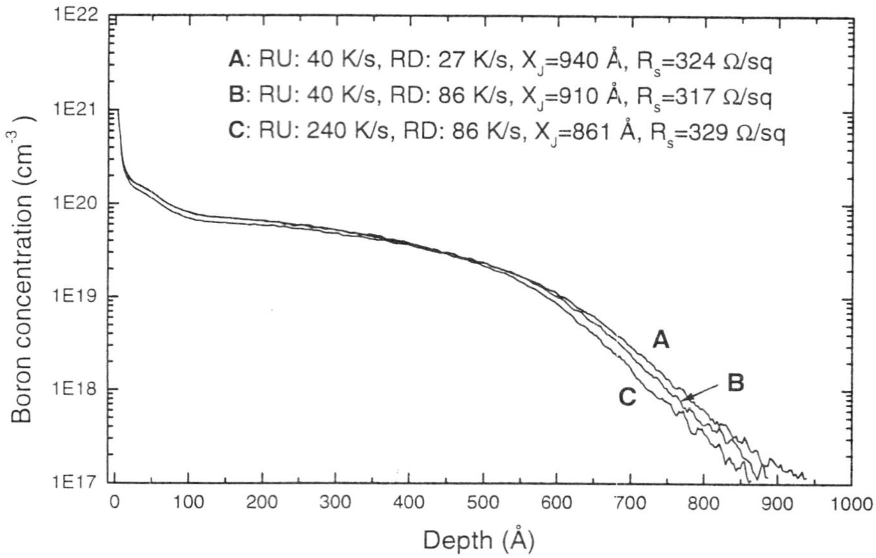

Fig. 4. SIMS profiles of $^{11}B^+$ 1 keV, $1 \cdot 10^{15} cm^{-2}$, 0° tilt, annealed at 1050°C 10 s without pre-stabilization at 750°C and two ramp-up rates of 40 K/s and 240 K/s and two different coolings, 86 K/s (extended) and 27 K/s (standard).

Fig. 4. The enhanced cooling has a stronger impact on the junction formation for the 40 K/s ramp-up (910 Å) compared to the 240 K/s on (861 Å) one. Transient activation and diffusion effects, well known from furnace technology, are generally reduced by a factor of three due to improved cooling. A further improvement is achieved by Spike Anneal Processing (SAP) of sub-10 keV implants. The system enables the highest ramp-up rates that are currently achieved in RTP. A characteristic spike anneal consists of a short ramp-up, no soak times at the elevated temperatures and fast cooling steps. Fig. 5 compares SAP results of fast and slow cooling with standard processing. A close inspection shows that a spike anneal with fast cooling yields a 200 Å shallower junction (740 Å) at 1050°C. This is in comparison to the standard process (940 Å) with prestab (40 K/s RU, 1050°C, 10 s, 27 K/s RD) only a modest increase in sheet resistance by 9%. The electrical junctions measured by Spreading Resistance Profiling (SRP) of both fast cooling processes with standard anneal (40 K/s RU) and the spike anneal (240 K/s RU) condition are determined to 720 Å and 521 Å, respectively, (see Fig. 6). All these results indicate that ultra-shallow junction formation can be improved by high ramp-up rates, fast cooling and, mostly important for reproducibility, the precisely controlled gaseous ambient. For deeper junctions, i.e. higher ion energies, the effects of high ramp-up rates and fast coolings minimize. Similar data of junction depth versus ramp-up rate and spike anneals with fast and standard cooling of 2.2 keV $1 \cdot 10^{15}$ cm^{-2} BF_2 implants are displayed in Figs. 7 and 8, respectively. A corresponding flash process for As annealing ($1 \cdot 10^{16}$ cm^{-2}, 20 keV) with a ramp-up rate of 200 K/s and a cooling-rate of 80 K/s resulted in a standard deviation of 0.53% (mean R_{sh} = 129.4 Ω/sq.) indicating a temperature variation of ±2°C. A 350 K/s ramp-up flash anneal showed a slightly higher sheet resistance (mean R_{sh} = 135.3 Ω/sq.) and standard deviation of 1.0%, indicating a slightly increased temperature variation of ±2.3°C.

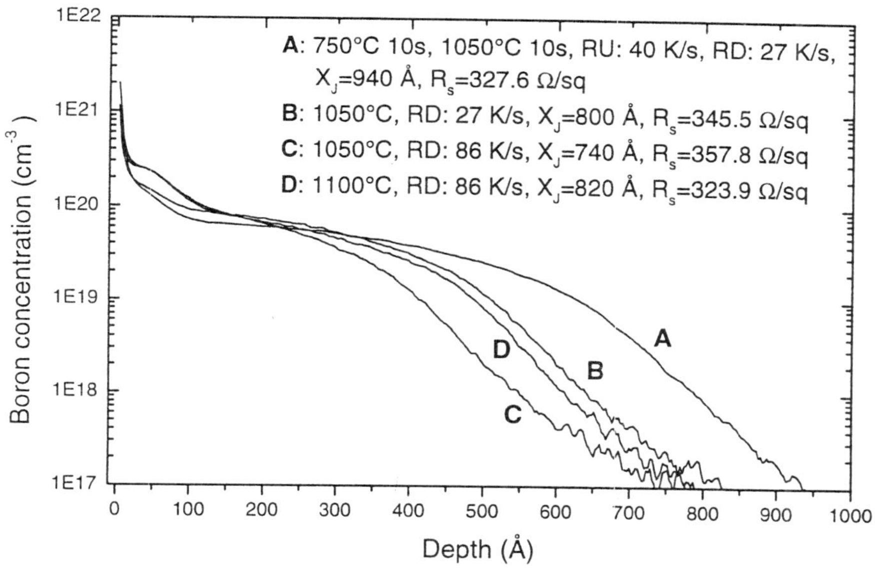

Fig. 5. SIMS profiles of $^{11}B^{+}$ 1keV, $1 \cdot 10^{15} cm^{-2}$, 0° tilt, annealed under standard conditions (pre-stabilization at 750°C 10 s, ramp-up with 40 K/s to 1050°C 10 s and a standard cooling of 27 K/s) compared to spike anneals (0 s soak time) up to 1050°C and 1100°C with fast and standard cooling.

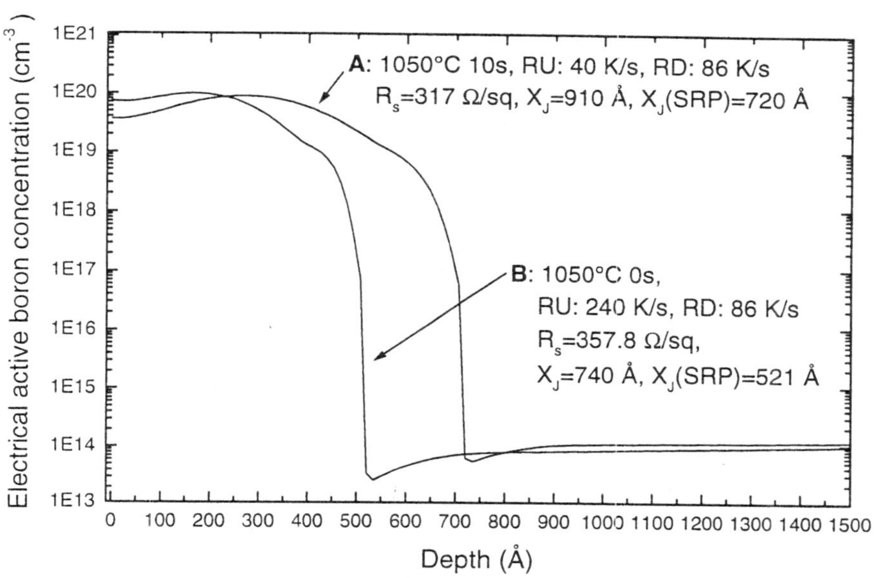

Fig. 6. SRP profiles of $^{11}B^{+}$ 1keV, $1 \cdot 10^{15} cm^{-2}$, 0° tilt, annealed at 1050°C for 10 s and a spike anneal at 1050°C with 240 K/s ramp-up rate. Both processes have an extended fast-cooling step and can be compared to the SIMS profiles shown in Fig. 5.

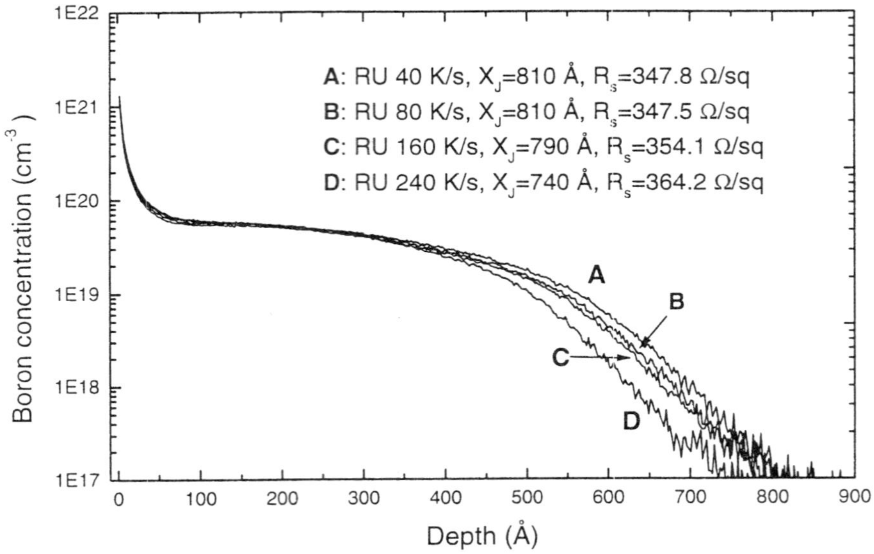

Fig. 7. SIMS profiles of $^{49}BF_2^+$ 2.2keV, $1 \cdot 10^{15} cm^{-2}$, 0° tilt, annealed at 1050°C 10 s without pre-stabilization at 750°C for 10 s and various ramp-up rates (40 K/s, 80 K/s, 160 K/s, 240 K/s) and extended fast-cool of 86 K/s.

5. Conclusions

A new RTP system characterized by a substantial improvement in system performance and temperature control has been presented. It features a unique process performance for different wafer

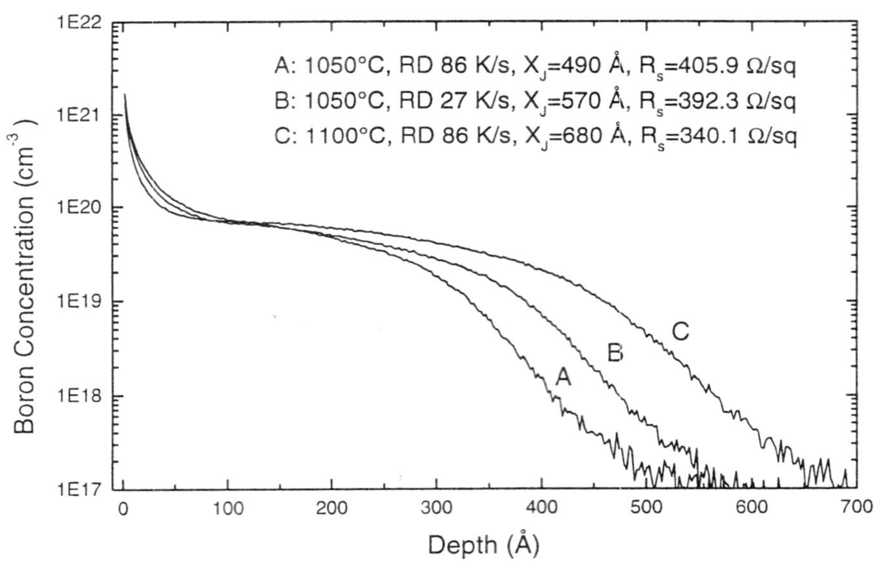

Fig. 8. SIMS profiles of $^{49}BF_2^+$ 2.2keV, $1 \cdot 10^{15} cm^{-2}$, 0° tilt, spike (0 s soak time) annealed with 240 K/s ramp-up rate up to 1050°C and 1100°C and standard (27 K/s) and extended fast-cool (86 K/s).

types in 200 and 300 mm operation meeting common and future process standards and requirements in a temperature range above 600°C. Best uniformities with standard deviations lower than 1% indicating temperature variations in a range of 2.4 K for 200 mm and 2.5 K for 300 mm wafers have been achieved. However, future processes will require a extension of the process window towards lower temperatures. Due to the dramatic change in wafer transmissivity below 600°C for medium and low doped wafers, extended evaluation techniques for reliable temperature control in the low temperature regime are necessary and currently under investigation. There is still room for further improvements.

Acknowledgements

The authors would like to thank A. Tillmann, Dr. Z. Nenyei and H. Walk for helpful discussions. The assistance and provision of data by D.F. Downey and J.W. Chow is gratefully appreciated. For further detailed information on ultra-shallow junction formation please refer to D.F. Downey et al.'s TMS '98 presentation [19]. Special thanks to the Applications Lab staff of STEAG AST Elektronik GmbH and co-authors of our previous communications.

References

[1] Dataquest, WSTS 8/95.
[2] ICE, Integr. Circ. Eng. Scot Scale, 1996.
[3] F. Roozeboom, in: F. Roozeboom (Ed.), Advances in rapid thermal and integrated processing, NATO ASI Series E, Vol. 318, Kluwer Academic Publishing, 1996.
[4] P. Singer, Thermal processing: The RTP race heats up, Semicond. Intern. 3 (1996) 64–68.
[5] A. Tillmann, U. Kreiser, P. Münzinger, S. Frigge, S. Buschbaum, P. Schmid, D. Löffelmacher, M. Merkwitz, T. Theilig, Transient thermal behaviour in a new RTP chamber, presented at 127th Ann. Meeting & Exhibition TMS'98, February 15–17, 1998, San Antonio, Texas, in press.
[6] A.T. Fiory, A.K. Nanda, Ripple pyrometry for rapid thermal annealing, Mat. Res. Soc. Symp. Proc. 342 (1994) 3–15.
[7] M. Oh, S.M. Merchant, B. Nguyenphu, Emissivity-independent control during phase transition in Rapid Thermal Processing, Proc. of 5th Int. Conf. on Advanced Thermal Processing RTP'97, pp. 199–202, 1997.
[8] M. Hauf, H. Balthasar, S. Müller, U. Müller, Emissivity compensated wafer temperature measurement using intensity-modulated lamp light, to be presented at 6th Int. Conf. on Advanced Thermal Processing RTP'98, Kyoto, Japan, 1998.
[9] Z. Nényei, H. Walk, T. Knarr, Defect-guarded Rapid Thermal Processing, J. Electrochem. Soc. 140 (6) (1993) 1728–1733.
[10] M. Glück, U. König, J. Hersener, Z. Nényei, A. Tillmann, Homogeneity of wet oxidation by RTP, Mat. Res. Soc. Symp. Proc. 342 (1994) 215–225.
[11] A. Tillmann, Model based temperature uniformity control during Rapid Thermal Processing, Proc. of 4th Int. Conf. on Advanced Thermal Processing RTP'96, pp. 356–371, 1996.
[12] A. Tillmann et al., Modeling and off-line optimization of a 300 mm rapid thermal processing system, to be published at E-MRS Spring Meeting, Symp. I, June 16–19, 1998, Straßbourg, France.
[13] Semiconductor Industry Association (SIA), National technology roadmap for semiconductors, Stevens Creek Blvd., Suite 271, San José, California, CA, 1997.
[14] S. Shishgushi, A. Minji, T. Hayashi, S. Saito, 1997 Symp. VLSI Dig. of Tech. Papers, p. 89.
[15] D.F. Downey, C.M. Osburn, S. Marcus, Ultra-shallow junction formation by ion implant and RTA, Solid State Technology 12/97 (1997) 71–82.

[16] W. Lerch, Ion implantation and rapid thermal annealing in synergy for shallow junction formation, Phys. Stat. Sol. (a) 158 (1996) 117–136.

[17] W. Lerch, M. Glück, N.A. Stollwijk, H. Walk, M. Schäfer, S.D. Marcus, D.F. Downey, J.W. Chow, H. Marquardt, Simulation of Rapid Thermal Annealed boron ultra-shallow junctions in inert and oxidizing ambient, MRS Spring Meeting April 13–17, 1998, San Francisco, California, to be published in Mat. Res. Soc. Symp. Proc., Vol. 525, 1998.

[18] S. D. Marcus, W. Lerch, D. F. Downey, S. Todorow, J.W. Chow, RTP Requirements to yield uniform and repeatable Ultra-Shallow Junctions with low energy Boron and BF_2 ion implants, presented at 127th Annual Meeting & Exhibition of the TMS 1998, February 15–17, 1998, San Antonio, Texas, to be published in J. of El. Mat. June/July (1998).

[19] D. F. Downey, S. D. Marcus, J.W. Chow, Optimization of RTP parameters to produce ultra-shallow highly-activated B^+, BF_2^+ and As^+ ion implanted junctions, presented at 127th Annual Meeting & Exhibition of the TMS 1998, February 15–17, 1998, San Antonio, Texas, to be published in J. of El. Mat. June/July (1998).

ELSEVIER

Microelectronic Engineering 45 (1999) 247–256

MICROELECTRONIC ENGINEERING

Novel process control strategies for 300 mm semiconductor production

Lothar Pfitzner[a,*], Richard Oechsner[a], Claus Schneider[a], Heiner Ryssel[a,b], Manfred Riemer[c], Mario von Podewils[c]

[a]Fraunhofer-IIS-B, Schottkystrasse 10, D-91058 Erlangen, Germany
[b]Lehrstuhl für Elektronische Bauelemente der Universität Erlangen-Nürnberg, Cauerstrasse 6, D-91058 Erlangen, Germany
[c]Thesys, Haarbergstrasse 61, D-99097 Erfurt, Germany

Abstract

Semiconductor industry is one of the fastest growing businesses. Integrated circuit production process is becoming more and more challenging as the increase in complexity and in wafer size continues. Semiconductor manufacturing comprises hundreds of mostly complex processing steps, almost each of them has to be controlled within a narrow process window. Measurement steps are more or less frequently used in the production processing. Quite often it is required to take those wafers that have to be tested or measured out of the production lots, which affects logistics and, even worse, adds costs for such monitor wafers. It is of utmost importance to check new strategies in monitoring for the enhancement of yield at any possible level. Three practical examples for integrated metrology, X-ray-induced photoelectron spectroscopy and ellipsometry integrated into a cluster tool and ellipsometry integrated into a vertical furnace have been demonstrated and will be described. Considerations of cost savings by integrated metrology are made concerning increase in equipment availability, yield improvement and reduction in cleanroom space.

1. Introduction

Semiconductor processing is driven by a steady increase in complexity of the microelectronic chips allowing higher number of gates and functions. This is achieved by a decrease of minimum dimensions and a multitude of additional number of processing steps. In contrast to these requirements, the production costs have to be reduced. Important contributions to such cost reductions arise from the use of larger wafers and from lower contamination and defect levels, combined with efforts towards better equipment utilization.

Above features typically require more monitoring and control steps. However, these non-productive steps are a significant cost factor. These additional steps also lower the cycle time, they require space in the cleanroom, and they require a high number of monitoring and test wafer. All this results in the

*Corresponding author. Tel.: +49-9131-761-110; fax: +49-9131-761-112.

0167-9317/99/$ – see front matter
PII: S0167-9317(99)00161-6

need of novel concepts and developments concerning less expensive advanced process control and monitoring strategies.

2. Integrated metrology

Monitoring has to be used for quality control in semiconductor production. Monitoring steps are performed to continuously control equipment state, process state, and wafer state. Monitoring is important during the ramp-up of a new fab, a new process, or a shrink version but also necessary for control of continuous operation.

In any case, monitoring of ramp-up and production is very costly and time-consuming. Therefore, new monitoring strategies must aim especially at a reduction in costs. This can be done by integrated metrology which is able to shorten the ramp-up periods as well as to reduce the amount of monitoring wafers. In this context, integrated metrology includes in situ metrology as well as integrated in-line metrology. In the case of in situ metrology, sensors are integrated into processing equipment and can be used for real-time data acquisition during the process and direct closed-loop process control. Integrated in-line metrology means that the sensor is directly attached to the processing equipment and offers immediate response to measurements and, optionally, feedback control.

In Fig. 1, three different scenarios for a fab ramp-up are shown: all curves start at a low yield assigned to the status when the equipment is inside the specifications. Curve one is the conventional learning curve. Curve two displays a faster ramp-up phase caused by a reduced cycle time due to the use of new monitoring techniques. A further increase in ramp-up can be achieved by minimizing the number of prototype runs. Especially prototypes are endangered to suffer from misprocessing. The integration of metrology may assist in such prototyping.

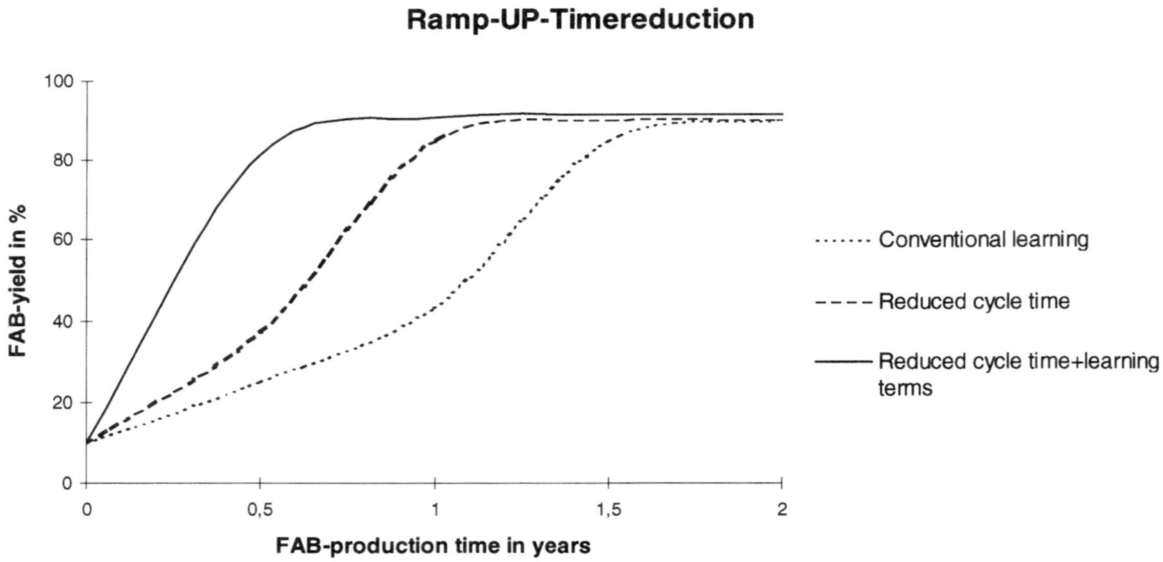

Fig. 1. Ramp-up time reduction for different strategies including integrated metrology.

In the following, three examples for integrated metrology are presented. Cluster tools offer possibilities to integrate metrology controlling the wafer state developed for in-line quality control as well as for implementation of feedback or feed forward control. In the first example described, X-ray-induced photoelectron spectroscopy (XPS) surface analysis was developed and integrated as an attached module [1]. Film thickness measurement by ellipsometry was integrated as a second example where integration was done in the form of in-line metrology. These monitoring tools allow to characterize cluster tools including cleaning steps by using XPS and the control of layer deposition

Fig. 2. Flexible MESC/CTMC compliant metrology module with XPS measurement system.

steps based on ellipsometry. The third example for integrated metrology consists of an in situ layer thickness sensor in a vertical furnace used for sensing layer growth and run-to-run control [2,3].

The XPS analysis was implemented in a separate proprietary module attached to the cluster platform due to throughput considerations and due to the need of an ultra-high vacuum [4]. The X-ray source and the detection system based on a cylindrical mirror analyzer are integrated as a single mechanical unit which allows a single-flange mounting (Fig. 2). UHV-compatible, high-precision *xy*-mapping and *z*-leveling for focusing are provided. The module uses its own module controller. A VME bus computer with a real-time, multi-tasking operation system OS-9 was employed. The software was developed by using process automation language (PAL) in a CONTROLVision environment (Brooks, Canada). The module controller supervises the states of the module components (XPS system, vacuum system and wafer mapping stage) and communicates with the cluster controller via a TCP/IP network (CTMC interface). The XPS system is fully controlled by the module controller. This includes the management of measurement recipes, the initialization of power supplies, the starting of measurement sequences as well as the calculation and the graphical representation of measurement results. For software and hardware integration of the attached module, the SEMI MESC/CTMC [5] standards were applied. Fig. 3 shows a comparison of measurements on the uncleaned wafer surface consisting of native oxide, the VPC-cleaned wafer surface and the regrown wafer surface. Different bonding states can be distinguished, thus allowing cleaning efficiency and surface passivation to be optimized and controlled [6].

A different approach was chosen for the integrated film thickness measurement. A commercial single-wavelength ellipsometer was mounted onto the transport chamber (Fig. 4). Since standard vacuum flanges were employed, a compact design was required. A proprietary optical light-guiding system was used to realize an arrangement of the sensor heads perpendicular to the platform. Film thickness measurements on the wafer, which reside on the robot end effector, can be performed routinely during the handling sequence with accuracy and repeatability close to the performance in stand-alone equipment (Fig. 5).

The third example for integrated metrology is an in situ layer thickness sensor in a vertical furnace

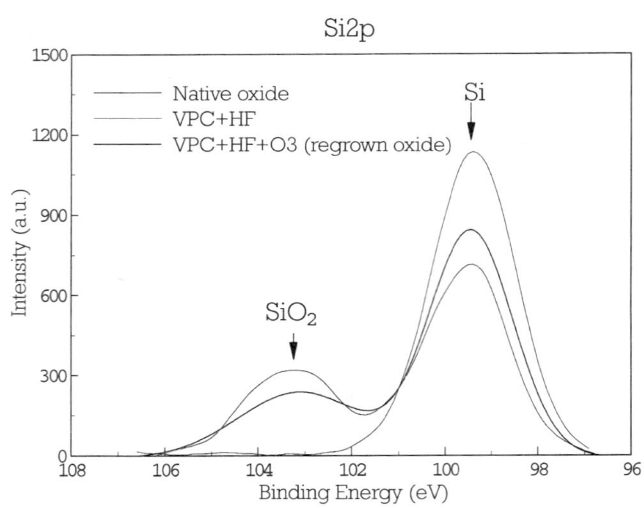

Fig. 3. XPS spectra for control of cleaning process in a gate stack cluster tool.

Fig. 4. Ellipsometer arrangement in the cluster platform.

used for sensing layer growth and run-to-run control [2,3]. Spectroscopic ellipsometry was chosen because of its high accuracy and versatility. The mechanical integration of the spectroscopic ellipsometer was performed only with minor modifications of the mechanical furnace geometry (Fig.

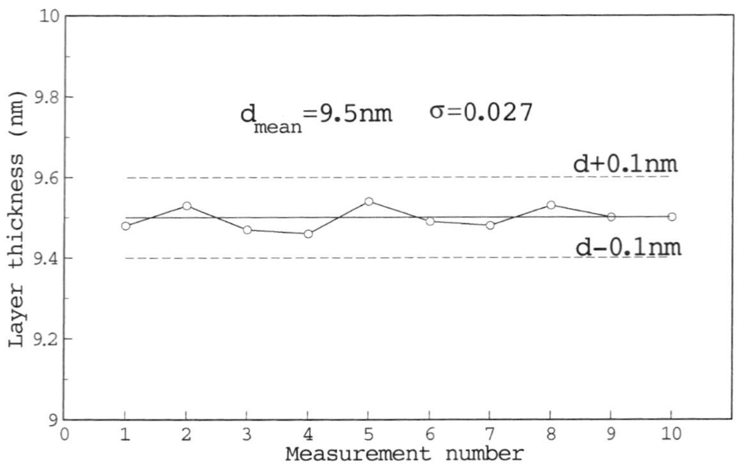

Fig. 5. Repeatability of layer thickness measurements using an ellipsometer integrated in a cluster tool. Between every measurement the wafer was placed into the load lock.

6). The two ellipsometer parts, the polarizer and the analyzer, were placed on the base plate of the furnace. Together with the base plate and the wafer carrier they form a mechanical unit which moves vertically with the boat loader. The ellipsometer light beam is guided by quartz glass prisms into the furnace tube and directed onto the wafer. The prisms are operated in total internal reflection mode. The introduced additional phase shift can be determined and subtracted from the measured phase shift. A real-time controller and a run-to-run controller integrated into the equipment controller utilize the layer thickness data measured in the in situ and in the post process mode for corrections during the process run and for the following run. Fig. 7 shows an example of a thermal dry oxidation process at 900°C which was terminated at an oxide thickness of 25 nm by end-point monitoring with the in situ spectroscopic ellipsometer. The oxide growth could not be stopped immediately by switching from oxygen to nitrogen flow. The oxide growth continued for an additional 0.35 nm. This post process

Fig. 6. Integrated ellipsometer in a vertical furnance.

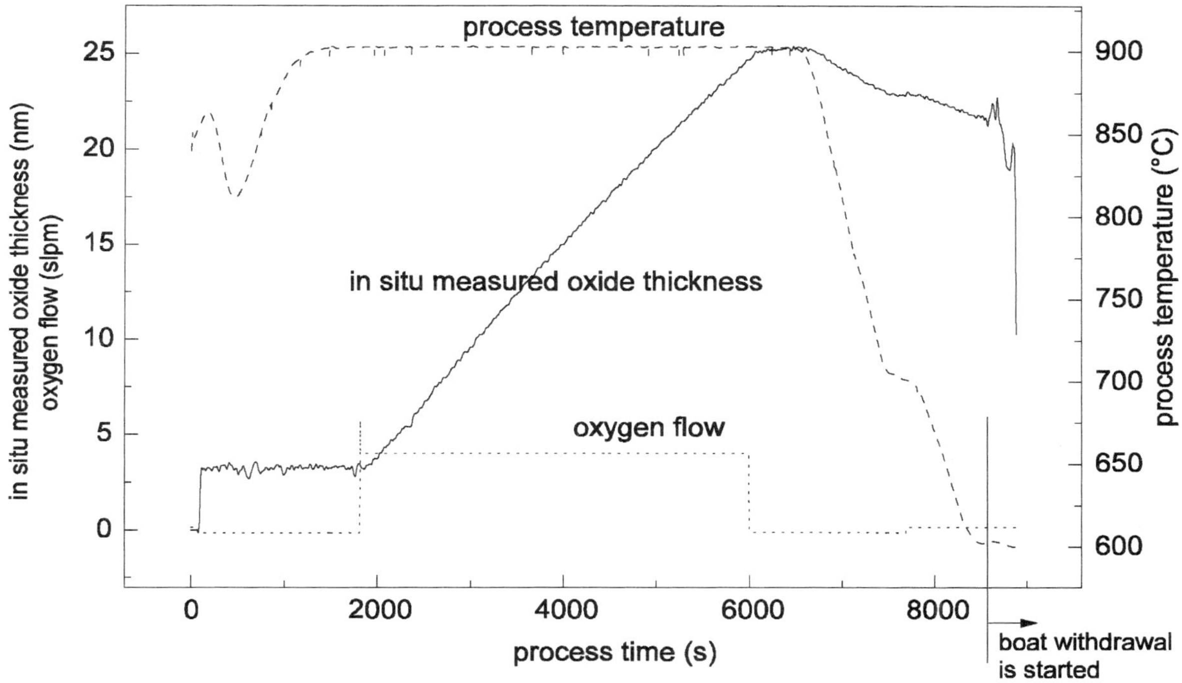

Fig. 7. Twenty-five nm thermal oxidation process at 900°C terminated by end-point-detection monitoring.

oxidation could easily be compensated. A precalculation by process simulation turned out to be the most flexible way, allowing an easy implementation into the computer for the process recipes.

3. Cost and yield considerations

The total productivity of a wafer fab depends on many factors. Equipment efficiency including throughput, availability, process performance, contamination, and defect generation is an important part of the cost-effectiveness of semiconductor production. Here, not only processing equipment, but also the metrology equipment is contributing and has to be taken into account [7]. Optimized work in progress (WIP) is another important cost factor. Additional benefits result from the reduction of cleanroom space, the reduction of wafer transport and handling operations, and the decrease in cycle times. For all the above-mentioned sources for enhancement of productivity, integrated metrology is of high importance. The benefit of integrated sensors will in many cases exceed the additional costs already within a few months.

Figs. 8–10 show estimations of typical cost savings by increase in equipment availability, yield improvement and reduction in cleanroom space. The shaded areas indicate the fields and the volume of possible cost reductions. A distinction was made with respect to estimations coming from mass production and from ASIC manufacture. In addition, mini-environments as a different approach for clean wafer environment are taken into consideration separately.

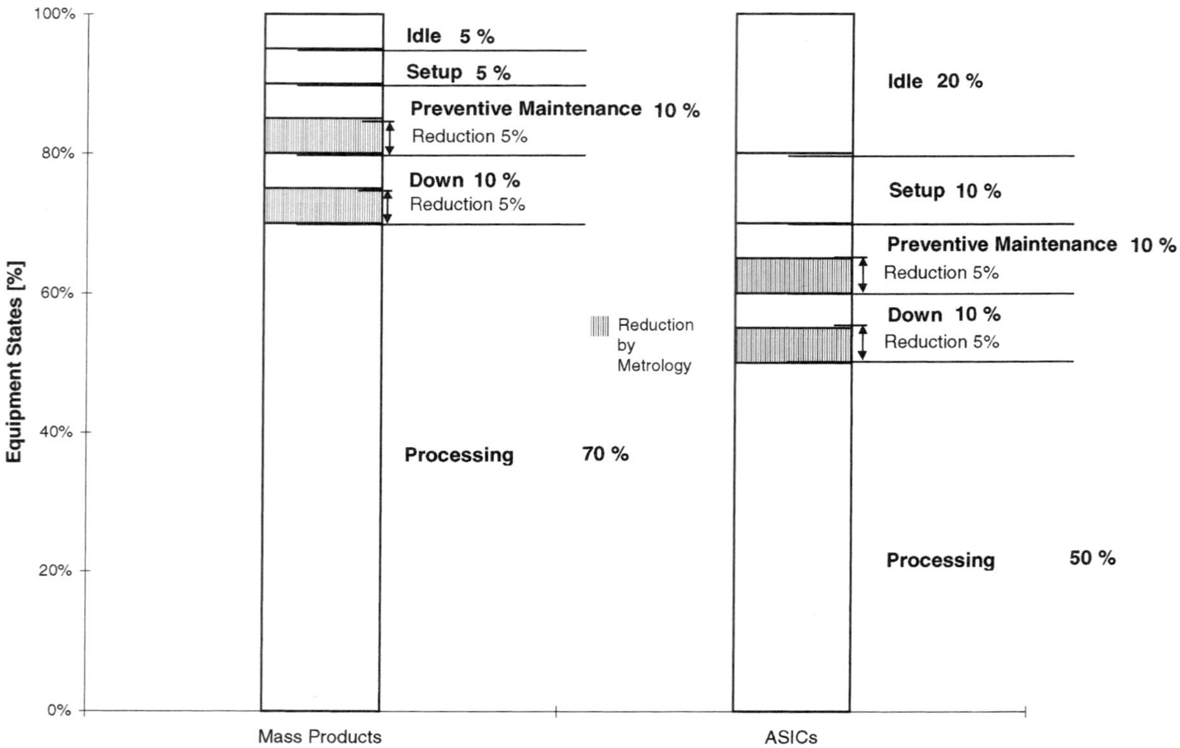

Fig. 8. Improvement of equipment availability by integrated metrology.

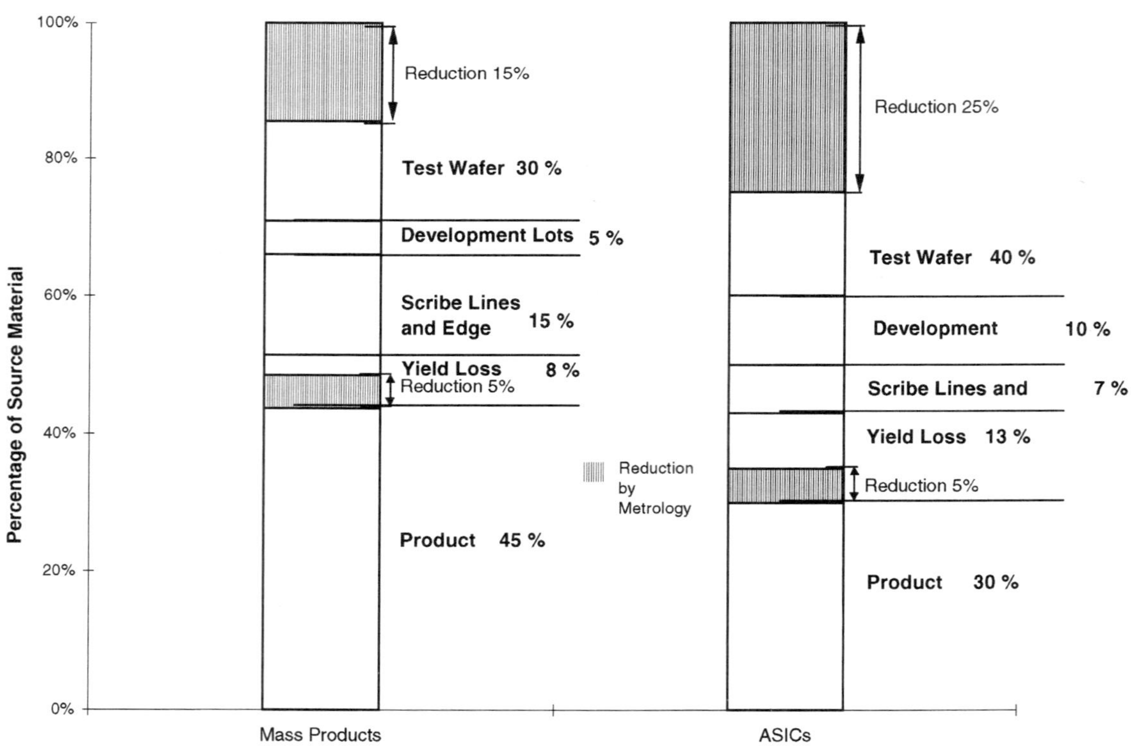

Fig. 9. Yield improvement by integrated metrology.

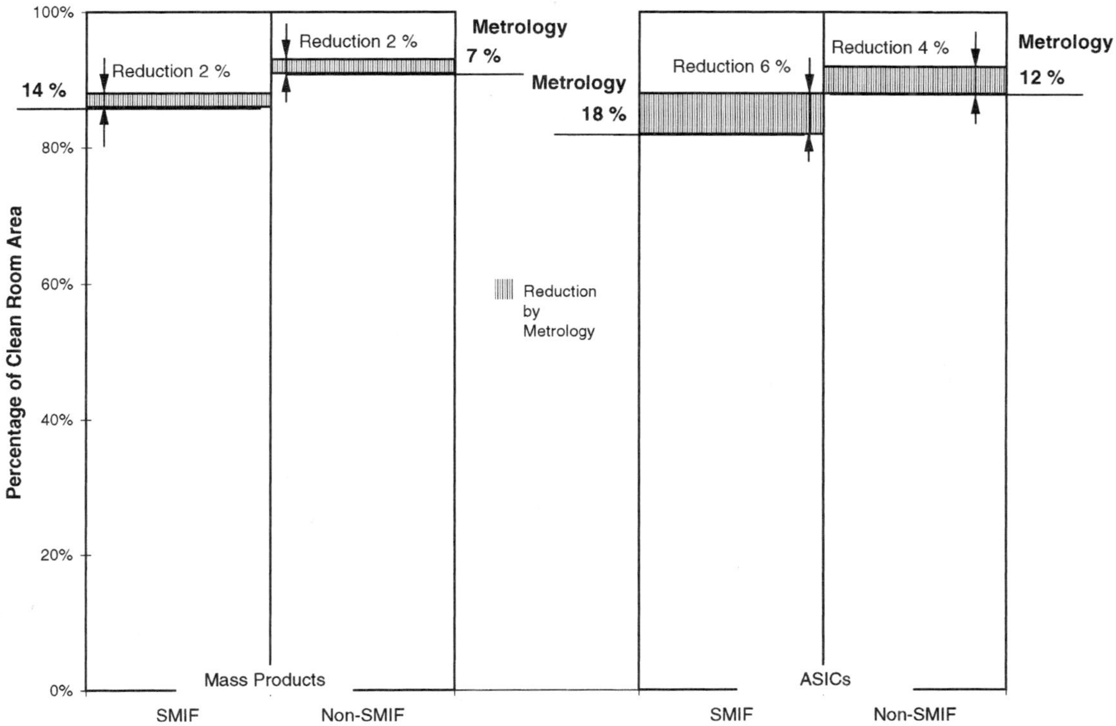

Fig. 10. Reduction of cleanroom space by integrated metrology.

4. Conclusions

Novel process control strategies will be based on increased application of integrated metrology. Such integrated metrology using in situ metrology or in-line metrology, whether based on existing equipment components or attached sensors/modules will be of high importance to reduce monitoring wafer, cleanroom space, learning turns. Especially, it will be of high importance to increase equipment availability and throughput. Reduction in costs and increase of productivity will be the most important results.

References

[1] I. Kasko, R. Öchsner, C. Schneider, L. Pfitzner, H. Ryssel, in: I. Olefjord, L. Nyborg, D. Briggs (Eds.), in ECASIA 1997 Proceedings, A Novel XPS System for Integration into Advanced Semiconductor Equipment for In-Line Process Control, Vol. 1089, Wiley, UK, 1997.
[2] R. Berger, C. Schneider, W. Lehnert, L. Pfitzner, H. Ryssel, in: SPIE-Proceedings on Process, Equipment, and Materials Control in Integrated Circuits Manufacturing II, Austin, Advanced Process Control System for Vertical Furnace, Vol. 2876, 1996, pp. 16–26.
[3] IVPS Final Report, Project No. 01M2925 B, Erlangen, 1996.

[4] L. Pfitzner, C. Schneider, H. Ryssel, C. Werkhoven, L. Deutschmann, O. Spindler, in: Proceedings of ISSM 96, Tokyo, The Challenge of Multi-Component, Multi-Vendor Cluster Tools, 1996, pp. 54–57.
[5] SEMI Book of Standards, Mountain View, 1996.
[6] B. Froeschle, F. Glowacki, A.J. Bauer, I. Kasko, R. Öchsner, C. Schneider, in: T.J. Riley, J.C. Gelpey, F. Roozeboom, S. Saito (Eds.), Mater. Res. Symp. Proc., Rapid Thermal and Integrated Processing VI, Cleaning Process Optimization in a Gate Oxide Cluster Tool Using an In-Line XPS Module, Vol. 470, 1997.
[7] C. Schneider, L. Pfitzner, H. Ryssel, in: Proceedings of ISSM 97, San Francisco, Modular Metrology Tools for Productivity Enhancement in Wafer Fabs, 1997, pp. B21–B24.

ELSEVIER

Microelectronic Engineering 45 (1999) 257–263

MICROELECTRONIC ENGINEERING

X-ray scattering from silicon surfaces: a useful tool for quality control

R. Stömmer[a,*], A.R. Martin[b], W. Hub[c], H. Göbel[a], U. Pietsch[d]

[a]*Siemens AG, Corporate Technology, Dept. ZT-MF 7, Otto-Hahn-Ring 6, D-81739 Munich, Germany*
[b]*Johann Wolfgang Goethe University, Institute for Anorganic Chemistry, Marie-Curie-Str. 11, D-60439 Frankfurt am Main, Germany*
[c]*Siemens AG, Semiconductors Division, Otto-Hahn-Ring 6, D-81739 Munich, Germany*
[d]*University of Potsdam, Institute of Physics, Am Neuen Palais 10, D-14469 Potsdam, Germany*

Abstract

The semiconductor industry started to conquer the nm-scale in process technology to create the next generation of microsized integrated circuits. With the down-scaling of semiconductor structures, surfaces and interfaces need precise control to achieve high yielding integrated circuits. For roughness determination of high quality surfaces and thickness control of ultra-thin layers, X-ray scattering proves to be an innovative tool. We demonstrate that specular and diffuse X-ray scattering from a silicon surface with a thin thermal oxide allows the precise determination of real structure parameters such as roughness, bulk density, oxide thickness, lateral correlation length and fractal dimension.

Keywords: Silicon surfaces; Semiconductor structures; Roughness; Bulk density; Oxide thickness; Lateral correlation length; Fractal dimension

1. Introduction

The semiconductor industry is increasing the wafer-diameter for integrated circuit (IC) fabrication from 200 mm to 300 mm. According to Moore's law, the geometry of semiconductor devices is further reduced. The design rule refers to the lateral gate length, which defines the technology generation. The gate-length is scaled down from 250 nm to 180 nm. Table 1 shows that the gate thickness of currently 4–5 nm needs further reduction, too. In consequence, the gate layer needs further perfection by reducing the interface and the surface roughness. The data of Table 1 are taken from the 'National Technology Roadmap for Semiconductors' [1]. Table 1 is completed with the roughness values provided by Wagner and Goodall [2,3].

For thin gate oxides with oxide thicknesses less than 10 nm, the Si/SiO_2 interface structure and the

*Corresponding author. Tel.: +49-89-6364-1411; fax: +49-89-6364-2256.
E-mail address: stoemmer@camelot.de (R. Stömmer)

0167-9317/99/$ – see front matter
PII: S0167-9317(99)00170-7

Table 1
Values from the roadmap, with focus on gate oxide thickness and surface/interface roughness. For each progress in technology, gate oxide thickness as well as roughness further decrease

Year	1997	1999	2001	2003	2006	2009	2012
Technology generation (nm)	250	180	150	130	100	70	50
Wafer diameter (mm)	200	300	300	300	300	450	450
Gate oxide thickness (nm)	4–5	3–4	2–3	2–3	1.5–2	<1.5	<1.0
Roughness (nm)	0.15[a]	0.1[a]	0.1[a]	TBD	TBD	TBD	TBD

[a] Roughness determined by atomic force microscopy; other methods of roughness determination may yield differing values [4–6]. TBD, to be determined.

SiO_2 surface structure affect the insulating function of the SiO_2 layer. In addition to that, the channel mobility of electrical charge carriers degrades with increasing Si/SiO_2 interface roughness [7]. Therefore, surfaces and interfaces have to be carefully monitored during wafer processing to keep them smooth [8]. X-ray scattering from surfaces and thin layers is a valuable tool to determine real structures on the nm-scale quantitatively and non-destructively. Its applicability to downsized semiconductor features of the near future is based on the weak interaction of sub-nm X-ray waves with matter. The sensitivity of X-rays to lateral and vertical density changes allows to investigate surface as well as interface and bulk properties of samples. In this way, the intensity and angular distribution of scattered X-rays allows to calculate surface and interface roughnesses, bulk densities, thicknesses of layers, and lateral properties such as the correlation length and the fractal dimension. According to the requirements of the roadmap [1], X-ray scattering seems among the most promising methods of future quality control [9].

2. Fractal surface parameters

For the Si sample under investigation we use the fractal description of surfaces [10]. Fig. 1 provides a surface profile that changes its appearance due to changing fractal parameters. The surface

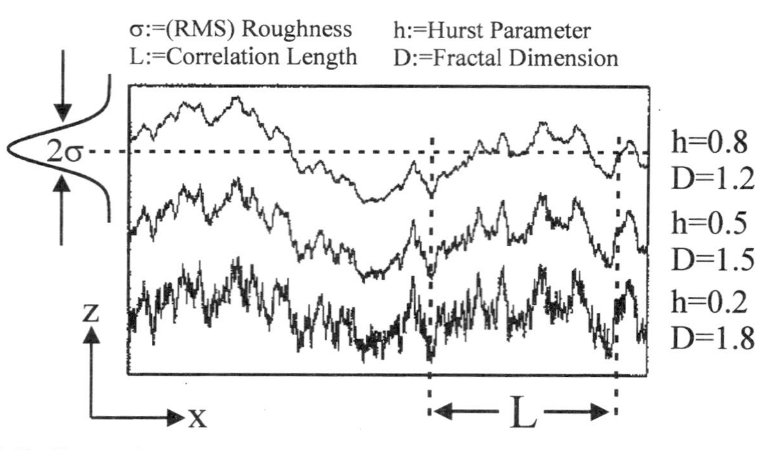

Fig. 1. Self-affine surface profiles can be described by the fractal parameters σ, L, h and D.

roughness σ denotes the half-width of the distribution of height fluctuations $(z - z^0)$ at the mean height z^0 of the surface. For the Si surface the distribution can be described as a Gaussian. The parameters L and h provide information on the lateral structure of surfaces. As can be seen in Fig. 1, the correlation length L is a measure for the length scale where the roughness σ does not change any more. Below L, the roughness σ decreases. Surfaces that show such a scaling behavior of roughness are called self-affine [11]. The Hurst parameter h can be understood by comparing the three surface profiles in Fig. 1, where σ and L are kept constant. A decrease of h can be described by an increase of high-frequency low-amplitude contributions to the surface profile. The decrease of h does not affect the roughness σ, although the surface profile looks more jagged for lower h. For such profiles, the Hurst parameter is related to the fractal dimension D by $D = 2 - h$. For surfaces in general, the fractal dimension D calculates to $D = 3 - h$. Surprisingly, but quite common for fractal objects, D is not an integer. D can be explained as a measure for the capacity of a fractal object to fill the space it is embedded in.

3. Instrumentation

The X-ray scattering experiment from the Si surface was accomplished at the reflectometer displayed in Fig. 2. It consists of an X-ray source, a curved, graded multilayer mirror on the incident beam side, a slit S_1 in front of the sample mounted onto the Ω circle of the goniometer, and two slits S_2, S_3 preceding the detector on the 2θ circle of the diffracted beam side. The X-ray source was a sealed 1.5 kW Cu tube providing an 8 mm$\times$0.04 mm line focus at a 6° take-off angle. The line focus was positioned in the focus of the parabolically curved, graded multilayer mirror (Göbel mirror) acting as beam condensor [12,13]. Additionally, the mirror separated the Cu K_{α_1} and Cu K_{α_2} lines for the scattering experiment. For small scattering angles, no further energy resolution was necessary [14]. The resulting wavelength was 0.154 nm. The purpose of the slit S_1 was to cut out a 0.2 mm$\times$15 mm part of the incident beam for the scattering experiment. By reducing the beam width from 1 mm to 0.2 mm by S_1, the intensity decreased by a factor of 5. This meant a loss for specular X-ray scattering, but it was necessary for the measurement and the interpretation of diffuse X-ray scattering. On the diffracted beam side, S_2 reduced the background scatter. S_2 was set to 0.4 mm$\times$15 mm and S_3

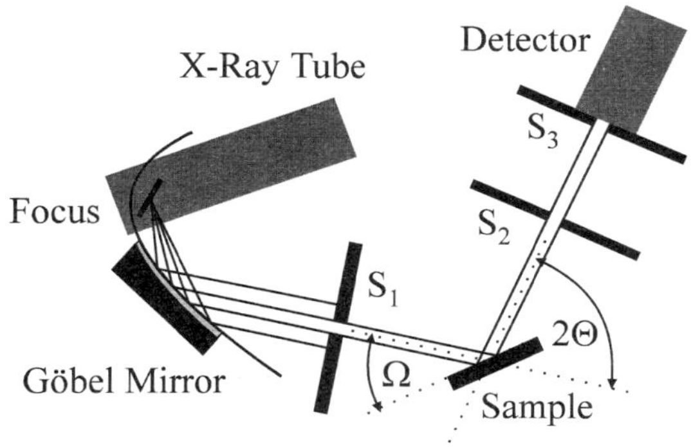

Fig. 2. The set-up of the reflectometer. The mirror attached to the X-ray tube acts as beam condensor for the scattering experiment.

was set to 0.3 mm×15 mm. With this set-up and an instrumental angular resolution of 0.06°, the primary beam intensity received at the scintillation counter detector was 10^8 cps. Eight orders of magnitude proved to be sufficient for the characterization of smooth Si surfaces with ultra-thin oxide layers.

4. Experiments, results and discussion

Specular X-ray scattering is sensitive normal to the sample surface. It provides vertical structure parameters, i.e. the density and the roughness of the Si surface as well as the density, the thickness and the roughness of the SiO_2 layers present on top of Si [15,16]. The inset in Fig. 3 displays the scattering geometry. By increasing Ω, the scattered signal is recorded by the detector at the angular position $\theta=\Omega$. Plotted on a logarithmic scale versus θ, the intensity distribution appears as a decreasing curve from $\theta=0°$ up to $\theta=3.5°$. All measured data are given as hollow squares, the simulations are displayed as lines. The exact shape of the specular signal is determined by the vertical structure parameters. To get information about the lateral structure of the surface, the diffuse scattering has to be measured [17]. Hereby, the angle between incoming beam and outgoing beam, 2θ, remains fixed. The angle between incoming beam and sample surface, Ω, is changed from $\Omega=0°$ to $\Omega=2\theta$. Two diffuse scattering curves are shown in Fig. 3 for $2\theta=1°$ and $2\theta=1.5°$. The sharp peaks in the middle of the diffuse scattering curves ($\Omega=\theta$) belong to the specular scattered intensity. The respective width of 0.03° is determined by the instrumental angular resolution of the setup. The specular signal has some distinct features: from $\theta=0°$ to $\theta=0.2°$ the intensity is nearly constant with a count rate of 10^8 cps. The critical angle of total external reflection θ_c at $\theta=0.226°$ allows to

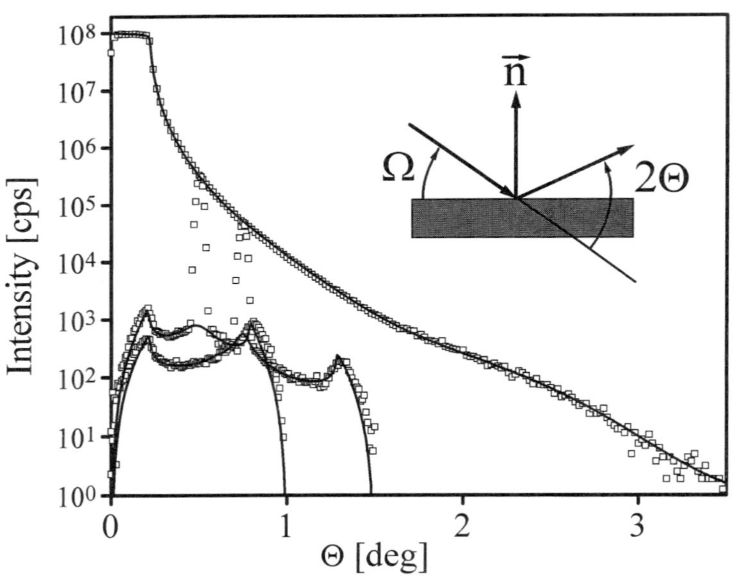

Fig. 3. Specular and diffuse X-ray scattering curves for the Si sample. The inset displays the scattering geometry.

recalculate the density of Si. The negative slope of the signal is affected by parameters such as the SiO_2 surface roughness, the Si/SiO_2 interface roughness, the density contrast between SiO_2 and Si, and the thickness of the SiO_2 layer. The thin SiO_2 layer causes a slight rise around $\theta = 2.4°$. The diffuse scattered signal originates solely in the imperfections of samples. With no surface or interface roughness, there would be no diffuse scattering. The tiny peaks on either side of the diffuse scattering curves in Fig. 3 originate from total external reflection. They are called Yoneda wings [18]. With the exception of the sharp specular peaks, the shape of the diffuse scattering curves is determined by the SiO_2 surface roughness and the Si/SiO_2 interface roughness, by the lateral correlation lengths and the fractal dimensions of the surface and the interface (see Section 2). The results obtained by the simulations in Fig. 3 are listed in Table 2. Additionally, the deviations from the mean values are provided.

The thickness value of the SiO_2 layer of 2.62 nm demonstrates the sensitivity of the X-ray scattering method to ultra-thin structures below 10 nm. An intensity of 10^8 cps is sufficient to quantify even thinner layers. The surface correlation length of 190 nm shows that the X-ray scattering method is well suited to characterize semiconductor structures on the nm-scale even laterally. A rather jagged surface profile can be inferred from the surface fractal dimension $D = 2.75$. The density value of the SiO_2 with 2.48 g/cm^3 is smaller than the oxide's theoretical density of 2.65 g/cm^3. This deviation may be attributed to the surface roughness of 0.4 nm, which is 15% of the SiO_2 thickness. The deviations from the mean values for the Si/SiO_2 interface parameters are considerably higher compared to the small deviations of the corresponding SiO_2 surface parameters. X-ray scattering is sensitive to the electron density contrast at interfaces. Therefore, the scattering process is dominated mainly by the air/SiO_2 interface, i.e. the SiO_2 surface. With 0.31 ± 0.09 nm, the interface roughness equals the surface roughness. From the interface correlation length of 410 nm, one can conclude that the interface profile is dominated by higher wavelength contributions compared to the surface profile. The appearance of the interface is as jagged as that of the surface, as the interface fractal dimension shows. For the Si sample under investigation, the X-ray intensity at our disposal had to cover at least eight orders of magnitude, as can be clearly seen in Fig. 3. The use of a Göbel-mirror attached to the X-ray tube proved to be very profitable.

Table 2
The results for the real structure parameters obtained from the simulations to the X-ray scattering data

Parameter	Result
SiO_2 surface roughness σ_{SiO_2}	0.40 ± 0.06 nm
SiO_2 layer thickness d	2.62 ± 0.15 nm
SiO_2 density ρ_{SiO_2}	2.48 ± 0.12 g/cm^3
SiO_2 surface Hurst parameter h	0.25 ± 0.06
SiO_2 surface fractal dimension D	2.75 ± 0.06
SiO_2 surface correlation length L	190 ± 25 nm
Si/SiO_2 interface roughness σ_{Si}	0.31 ± 0.09 nm
Si density ρ_{Si}	2.32 ± 0.05 g/cm^3
Si/SiO_2 interface Hurst parameter h	0.3 ± 0.15
Si/SiO_2 interface fractal dimension D	2.7 ± 0.15
Si/SiO_2 interface correlation length L	410 ± 80 nm

5. Conclusion

There is an urgent need for new methods of quality control to keep pace with the continuous downscaling and improvement of semiconductor structures. We have shown for Si wafers with an ultra-thin SiO_2 layer on top that X-ray scattering is a promising candidate for future quality control in the front end processes of IC production. Obviously, X-ray scattering is not restricted to silicon. The method is sensitive to density contrasts and therefore it can be applied to all sorts of unpatterned surfaces and layered structures [19–22]. X-ray scattering is non-destructive. It is applicable to structures on the nm-scale and roughnesses on the sub-nm-scale. Due to its sensitivity for density fluctuations, X-ray scattering reveals the inner structure of ultra-thin layers. For precise structure determination, reflectometers need a high flux of X-rays and a good angular resolution. Therefore, we applied a Göbel-mirror as beam condensor. The reflectometer set-up has to be designed in a compact and maintenance-free fashion for use in IC fabrication. Additionally, the wafer-handling prior to and after X-ray measurements needs automation, and the software for data evaluation needs to be adjusted to the specific processing step which has to be controlled. With respect to the continuous shrinking of device geometry as documented in Table 1, and the fact that the development of fully automated, high flux reflectometers takes at least 2 years, technologists of metrology departments should focus their attention onto X-ray scattering now.

Acknowledgements

The fruitful discussions with Alexander Olbrich (Siemens Semiconductor Division, Munich, Germany) and with Alexander Ulyanenkov (RIGAKU Corporation, Tokyo, Japan) about the properties of thin oxides are highly appreciated.

References

[1] The National Technology Roadmap for Semiconductors 97, published by the Semiconductor Industry Association, 1997.
[2] P. Wagner, H.A. Gerber, R. Schmolke, R. Velten, SPIE 2862 (1996) 152.
[3] R.K. Goodall, SPIE 2862 (1996) 2.
[4] P. Wagner, H.A. Gerber, D. Gräf, R. Schmolke, M. Suhren, SPIE 2862 (1996) 18.
[5] B.W. Scheer, SPIE 2862 (1996) 78.
[6] R. Stömmer, A.R. Martin, Th. Geue, H. Göbel, W. Hub, U. Pietsch, Adv. X-ray Anal. (1998) in press.
[7] T. Ohmi, K. Kotani, A. Teramoto, M. Miyashita, IEEE Trans. Electron. Dev. Lett. 12 (1991) 652.
[8] G. Kamarinos, P. Felix, J. Phys. D: Appl. Phys. 29 (1996) 487.
[9] M. Dax, Semicond. Int. (1996) 91, August.
[10] J.-F. Gouyet, M. Rosso, B. Sapoval, in: Fractals and Disordered Systems, Springer-Verlag, 1991, p. 229.
[11] B.B. Mandelbrot, The Fractal Geometry of Nature, Freeman, New York, 1982.
[12] M. Schuster, H. Göbel, J. Phys. D: Appl. Phys. 28 (1995) A270.
[13] M. Schuster, H. Göbel, J. Phys. D: Appl. Phys. 29 (1996) 1677.
[14] J.D. Shindler, R.M. Suter, Rev. Sci. Instrum. 63 (1992) 5343.
[15] L. Nevot, P. Croce, Rev. Phys. Appl. 15 (1980) 761.
[16] R.A. Cowley, T.W. Ryan, J. Phys. D: Appl. Phys. 20 (1987) 61.
[17] S.K. Sinha, E.B. Sirota, S. Garoff, H.B. Stanley, Phys. Rev. B 38 (1988) 2297.

[18] Y. Yoneda, Phys. Rev. 131 (1963) 2010.
[19] S. Banerjee, M.K. Sanyal, A. Datta, S. Kanakaraju, S. Mohan, Phys. Rev. B 54 (1996) 16377.
[20] R. Stömmer, J. Grenzer, J. Fischer, U. Pietsch, J. Phys. D: Appl. Phys. 28 (1995) A216.
[21] W.F.J. Slijkerman, J.M. Gay, P.M. Zagwijn, J.F. van der Veen, J.E. MacDonald, A.A. Williams, D.J. Gravesteijn, G.F.A. van de Walle, J. Appl. Phys. 68 (1990) 5105.
[22] S.C. Woronick, B.X. Yang, A. Krol, Y.H. Kao, H. Munekata, L.L. Chang, J. Appl. Phys. 66 (1989) 3566.

ELSEVIER

Microelectronic Engineering 45 (1999) 265–268

MICROELECTRONIC ENGINEERING

An industrial SR-TXRF facility at ESRF

F. Comin*, P. Mangiagalli, M. Navizet, G. Apostolo

European Synchrotron Radiation Facility (ESRF), BP 220, 38043 Grenoble, France

Abstract

A TXRF industrial facility for the mapping of trace impurities on the surface of 300-mm Silicon wafers is presently in the construction phase and will start the commissioning phase at the end of 1998. The elements to be detected range from Na to Hg with a target routine detection limit of 10^8 at/cm^2 and the capability of mapping the surface of 300-mm wafer with a resolution of 500 pixels and a throughput of three wafers/h. © 1999 Elsevier Science B.V. All rights reserved.

Keywords: TXRF; Silicon wafers; Metallic contamination

1. Introduction

Metallic contaminants at the surface of silicon wafer heavily hinder the performances of integrated circuits. The present IC technology of 0.25 μm design rule on 200-mm wafer can tolerate contaminant concentrations between 2 and 5×10^{10} at/cm^2 that in turn requires a maximum surface detection limits between 2 and 5×10^9 at/cm^2. At this level of ultimate detection the existing non-destructive method for measuring metallic surface contamination, total reflection X-ray fluorescence (TXRF) approaches its limits. The National Technology Roadmap for Semiconductor published by SEMATECH[1] has already forecasted an important leap for the TXRF technique in order to cope with the new industrial requirements, namely the moving from laboratory-based apparata to centralised synchrotron radiation facilities. In these facilities the 10 orders of magnitudes of increase in brilliance of the X-ray beams can well be exploited to meet the future requirements of IC manufacturer and provide new possibilities in wafer TXRF: mapping capabilities and low Z atom detection. The advantages of synchrotron beams over the isotropic emission of laboratory sources stem from a variety of factors ranging from the bare intensity concept, the high collimation, large energy tunability and high linear polarisation. Each of these factors carries a specific advantage that is kept accounted for and integrated in a TXRF facility that is presently in the construction phase at the European Synchrotron Radiation Facility (ESRF). The facility follows-up previous studies carried out in other synchrotron radiation centres [1–6]. A dedicated pilot experiment has been carried out at ESRF [7] to

*Corresponding author. Tel.: +33-4-7688-2000; fax: +33-4-7688-2020.

[1](http://notes.sematech.org/1997mem.htm)

0167-9317/99/$ – see front matter
PII: S0167-9317(99)00192-6

verify the possibility of adopting, at variant with the previous experiments, a convenient horizontal geometry for the wafers while keeping the detector aligned with the polarisation vector.

1.1. Aims of the facility

The ESRF-TXRF facility will provide silicon manufacturing companies with an analytical centralised TXRF service capable either of ultimate detection limits below 10^8 at/cm^2, or with the possibility of mapping the concentration of contaminants with an initial detection limit of 10^9 at/cm^2 for 1999 to be pushed down to 10^8 at/cm^2 in the following years. The wafer size can be either 200 or 300 mm and the pixel size of about 1 cm^2. The project envisages also the detection of low Z elements down at least to sodium, thus requiring excitation beams below the silicon K threshold.

2. Facility layout

2.1. Optics

The key issues that have been taken in account for the definition of the optical layout of the installation can be summarised as follows:

- Need of a wide energy range from 1 to 40 keV for optimising the detection sensitivity for every element. The low energy range below the Si threshold is necessary for the low Z atoms detection.
- Need to reject as much as possible the harmonic content of the X-ray beam to limit detection problems with standard solid-state energy dispersive detectors.
- Need of preserving the polarisation and collimation of the beam for maximum rejection of the elastically scattered radiation at the detector.
- Need of a monochromator with fixed exit beam to avoid the need to realign the wafer position at each change of energy.

2.1.1. The photon source

The best way to reject harmonics is, of course to start without harmonics. We asked the Insertion device group of ESRF to analyse the possibility of manufacturing an aperiodic undulator which characteristics is to have the ‘harmonics’ well displaced from where they should be if they were normal harmonics. From their calculation [9] the residual harmonic content of the undulator that is presently in construction is a few percent. A second reduction of the harmonic content can be achieved at the monochromator.

2.1.2. The monochromator

For maximum reliability the X-ray monochromator is based on a single rotation motion following well established concepts. The dispersing elements are a silicon (111) channel cut for the energy range 2.8–40 keV and a multilayer pair for the low energy. The high brilliance of ESRF leave the possibility of using the Si(111) reflection without degradation of performance. This allows to perform chemical recognition by collecting XANES spectra around the threshold of the atomic species of interest. The detailed geometry of the monochromator is described in Ref. [8]. Because of the need of reaching energies as low as 1 keV the beamline has no window along the 40 m that separate the wafer from the storage ring and special care is taken to avoid mutual cross-contamination and vacuum accidents.

2.2. The TXRF set-up

The TXRF end station is embedded in a class 100 laminar flow hood located in a shielded hutch of about 25 m^2. The station encompasses a loading robot that transfer the wafers from standard cassettes to a load-lock; a high vacuum handler that transfers the wafers from the load-lock to the measuring position in the main TXRF chamber, a set of two seven-element linear arrays of Si(Li) detectors for the acquisition of the fluorescence spectra, and a wafer chuck/manipulator for the correct positioning of the wafer onto the X-ray beam and for rotating the wafer around his vertical axis in order to perform the impurity mapping. The 14 channels of the detector are split in two independent arrays for servicing/reliability purposes.

Due to the grazing incidence, the beam illuminate a strip on the sample, the use of 14 independent detector channels each of whom looks to a 10.5-mm wide portion of this strip allows an easy mapping of the entire surface of the wafer just by rotating the sample and collecting data in parallel from the detector's array.

This geometry strongly simplifies the mechanical requirements for the wafer handling, but implies that either the spatial resolution or the sensitivity for the outer portion of the wafer is somewhat lower that for the central one.

2.3. Time schedule

The facility will start its commissioning phase early 1999 with the first tests on real 300-mm wafer. This will lead to a field assessment of the specifications attained and further tests are already envisaged to ascertain which other opportunities can be exploited with the instrument. Two directions seems particularly interesting, namely the possibility of performing XANES spectroscopy in selected atomic species for a chemical identification of the contaminant and detection of carbon species via electron spectroscopic techniques.

Acknowledgements

This work has been carried out through a grant from the MEDEA program Package T618. The authors gratefully acknowledge Erick Ziegler for the prompt preparation and characterization of the multilayers

References

[1] A. Iida, Adv. X-Ray Anal. 35 (1992) 795.
[2] R. Rieder et al., Nucl. Instrum. Methods Phys. Res. A355 (1995) 648–663.

[3] P. Pianetta et al., Rev. Sci. Instrum. 66 (1995) 1293.
[4] R. Goergl et al., X-Ray Spectrosc. 26 (1997) 189–194.
[5] P. Wobrauschek et al., Spectrochim. Acta Part B 52 (1997) 901–906.
[6] C. Streli et al., Nucl. Instrum. Methods Phys. Res. A345 (1994) 399–403.
[7] L. Ortega et al., J. Synchrotron Radiat. 5 (1998) 1064–1066.
[8] F. Comin et al., SPIE Proc. Int. Symp. Optical Sci., Eng., Instrum., San Diego, July 1998, SPIE Proceedings 3448 (1998) 11–18.
[9] J. Chavanne et al., Proc. EPAC Conf., Stockholm, June 98, IOP Publishing.

ELSEVIER

Microelectronic Engineering 45 (1999) 269–276

MICROELECTRONIC ENGINEERING

SOPRA SE300: a new tool for high accuracy characterization of multilayer structures

Pierre Boher*, Marc Bucchia, Jean Pierre Rey, Jean Louis Stehle

SOPRA S.A., 26 rue Pierre Joigneaux, 92270 Bois-Colombes, France

Abstract

In order to characterize 300-mm wafers at different stages of the IC manufacturing, a new tool based on spectroscopic ellipsometry has been recently developed at SOPRA. This new instrument called SE-300 has some important new features compared to the other ellipsometers of SOPRA or of the competition. First the optical setup allows to obtained very small measurement spots down to 35 × 45 μm in polychromatic light to be able to work from deep UV 190 nm to near infrared; second the combined monochromator/spectrometer is directly setup on the analyser arm and allows both multichannel and scanning measurements on the same spot. Scanning measurement made with a real double monochromator including prism and grating allows very accurate measurement that can be used to extract optical indices and solve complex multilayer structures. Multichannel measurements are made through a prism/grating spectrometer with quasi-linear dispersion in wavelength. All the elements (robotics, prealigner, *XYZ* mapping, pattern recognition) are fully compatible with the new generation of 300-mm wafers. Practical results obtained in a realistic environment (I300I acceptance tests and MEDEA project) are presented.

1. Introduction

The evolution of integrated circuits is now for more than 25 years in the direction of higher speed and greater complexity, with higher levels of integration and lower power consumption. The new generation of 300-mm wafer technology is based on planar critical dimensions in the range 0.2–0.3 μm with the thinnest layer thicknesses in the range 20–40 Å. These processes can have ten or more different layers of material in one structure. Another aspect of this new technology is the suppression of test wafers mainly for economical reasons. So, the requirements in terms of non-destructive characterization become more and more critical since the measurement must be made directly on the production wafer without contamination and with an unprecedent accuracy.

Spectroscopic ellipsometry has long been recognized as a powerful technique to characterize thin films and complex multilayers. SOPRA commercialized the first spectroscopic ellipsometer in 1983. Since this date, numerous systems have been sold for research and development and the technique

*Corresponding author. Tel.: +33-1-4781-0949; fax +33-1-4242-2934.
E-mail address: sopra-sa@sopra-sa.com (P. Boher)

0167-9317/99/$ – see front matter
PII: S0167-9317(99)00173-2

became well established for microelectronics characterization. In 1986, SOPRA first developed a new instrument including a spectrograph and a silicon array photodetector [1] opening the door to rapid spectroscopic measurement compatible with in-line requirements of the semiconductor industry. In 1992 SOPRA proposed a new tool, the multilayer monitor (MLM) dedicated to in-line applications [2]. This instrument designed for 100–200-mm wafers has been sold in the majority of production facility around the world. Thank to its high quality measurements which allow in particular extraction of optical indices from scanning measurements it was rapidly seen as a reference in the field and a way to calibrate other less sophisticated instruments from the competition. The main limitations of this instrument were its quite large spot size (100×100 μm) and the impossibility to work in the deep UV 190 nm wavelength range. For real time process control, SOPRA developed also in 1996 a real-time spectroscopic ellipsometer (RTSE) for easy adaptation on deposition chambers [3–5]. This new instrument was designed for speed and measurement accuracy with an intensifier just before the multichannel detector. The main advantages were a best signal-to-noise ratio and a best efficiency in the UV range.

In 1997, with the occurrence of the new generation of 300-mm wafers, SOPRA took the opportunity to develop a completely new on-line system with advanced performances in terms of spot size and spectral range. This new system is described here with some results obtained during its successful gauge study performed at the International 300-mm Initiative (I300I) [6].

2. Description of the SE300

2.1. Optical design of the system

This is an improved version of the optical setup of the TENCOR UV-1250SE instrument designed by SOPRA [7] (cf. Fig. 1). The light source is a high stability 75 W xenon arc lamp, which is placed directly in front of the first aperture to maximize the signal. The source size is defined by the aperture of the source slit, which is typically 500×100 μm. The focusing mirror is a 10 mm focal length Al coated ellipsoidal mirror designed to minimize imaging and polarization defects. The magnification factor is about X37.5. The theoretical width of the spot in X direction (perpendicularly to the plane of incidence) without diffraction, mirror and alignment defects is consequently $500/37.5 = 13.3$ μm. In Y direction, (along the plane of incidence) the large angle of incidence spreads the spot by a factor of 2–4 depending on the angle of incidence. The length of the spot in Y direction at 75° is about 11 μm. The 500×100 μm source aperture gives approximately a square spot on the sample. Due to the large numerical aperture (0.25) of the light focused on the sample the diffraction pattern is reduced to about 2 μm at a wavelength of 0.5 μm. After reflection on the sample the light is collected by a second mirror which focused the light through the analyzer on the entrance of the spectrograph. The angle of incidence is selected by a slit placed in front of the collecting mirror. The angle of incidence must be selected after the light is reflected from the sample otherwise the numerical aperture of the focused beam is reduced, which yields to the increase of the diffraction pattern on the sample multiplied by the projection angle [8]. The slit width defines the beam angular divergence (typically $\pm 1°$) and its position gives the angle of incidence. A second rochon polarizer is used to analyze the polarization state of the light reflected from the sample. The spectrograph has two separated outputs. One is coupled to an intensified photodiode array (IPDA) for fast data acquisition and the second one is

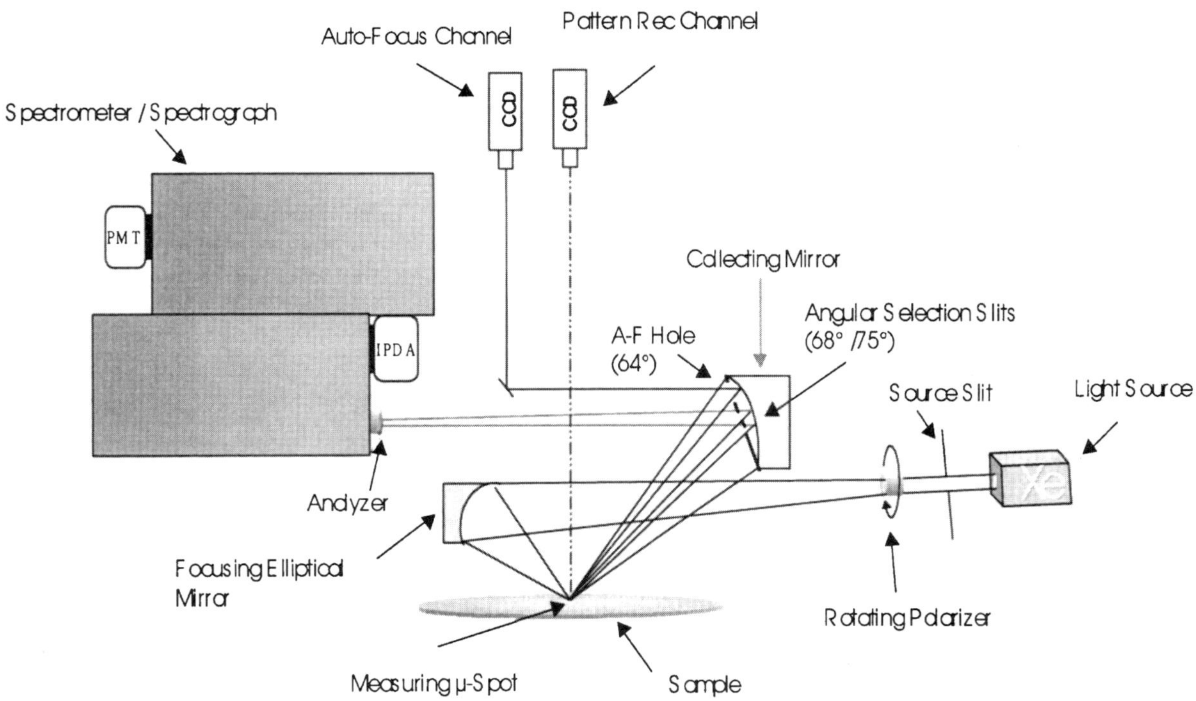

Fig. 1. Optical setup of the SE300 system from SOPRA.

coupled to a PM Tube for scanning a wide spectral range, high resolution, high signal-to-noise and high accuracy. All optics have been designed to ensure high UV efficiency and minimum aberrations in the focused beam. The spectral range covered by this ellipsometer is 190–850 nm in scanning mode and 240–840 nm in multichannel mode.

2.2. Microspot characteristics

The ability to measure in a small box depends not only on the optical setup, but also on the possibility to adjust the measurement beam in the right position. In the case of the SE300 the accuracy off the $X-Y$ table is about 5 μm. The beam size can be reduced down to 20 μm adjusting the source slit, but the signal intensity becomes very small and the measurement is difficult. In fact practical beam size is around 40×40 μm (cf. Fig. 2). Normal beam size conditions for optimized signal-to-noise ratio are 70×70 μm.

2.3. Software and throughput

The system is completely computer controlled with a new generation of user friendly software under Windows NT™. Different modes are available for the user: the research mode is used to extract new optical indices and define strategies to analyse new multilayer structures; the engineer mode

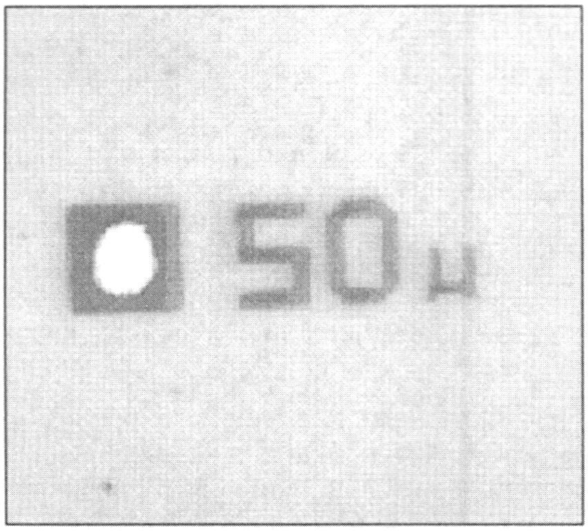

Fig. 2. Measurement spot in a 50 × 50-μm box.

allows to define measurement recipes and adjust the structural models for automatic regressions; the operator mode makes automatically a measurement procedure on one or more wafers with automatic fitting and display of the results. Communication with the host computer can be made via SECS II protocol [9]. The throughput for five measurement points and three parameters by points are 24 and 97 wafers per hour for the scanning and multichannel mode, respectively.

Table 1
Comparison of results obtained on different kind of samples with two SOPRA instruments: the ES4G reference ellipsometer and the SE300 system in scanning and multichannel mode

Sample type	Parameter (thickness, nm)	SE300 scanning	SE300 multichannel	ES4G scanning
SiO_2/Si native	SiO_2	3.82±0.15	4.03±0.1	4.13±0.1
SiO_2/Si reference	SiO_2	123.9±0.3	124.4±0.3	123.6±0.3
SiO_2/Si thick	SiO_2	1996.8±0.2	1998.4±0.4	1991.5±0.1
SiN_x/Si	SiN_x	152.9±0.2	153.3±0.3	152.4±0.1
$SiN_x/SiO_x/Si$	SiN_x	6.4±0.3	6.7±0.4	7.5±0.2
	SiO_2	2.9±0.2	2.8±0.5	1.4±0.2
OPO	SiO_2	3.6±0.6	3.9±1	3.7±1
	Poly	150.6±0.2	150.7±0.3	150.5±0.2
	SiO_2	115.5±1	Fixed 115.5	111.6±1
ONOPO	SiO_2	7.5±0.4	11.8±1	7.5±0.6
	SiN_x	16.1±0.6	13.6±0.6	16.5±0.3
	SiO_2	2.5±0.3	2.7±0.7	2.1±0.4
	Poly	151.2±0.3	151.5±1	150.8±0.3
	SiO_2	112.6±2	Fixed 112.6	110.1±2

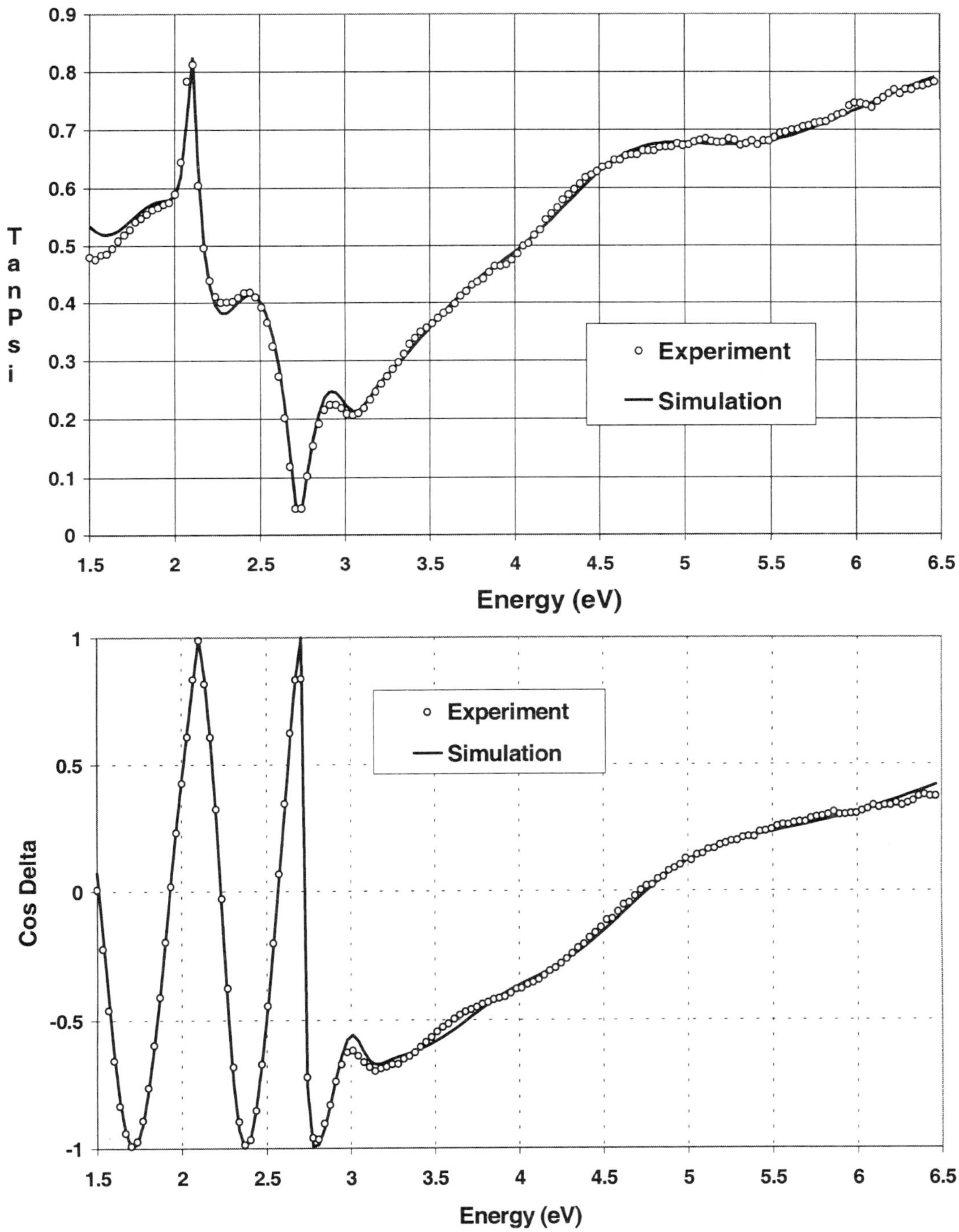

Fig. 3. Experimental and simulated SE measurements made on a OPO sample using the SE300 system in scanning mode.

3. Experimental results

3.1. Performance for different types of samples

Different samples characteristic of the microelectronic technology have been measured with the SE300 system in scanning and IPDA mode, and with the SOPRA ES4G system which has been recognized for many years as a reference. This instrument is in fact routinely used at VLSI standards for the same kind of analysis [10]. Seven samples have been measured, three SiO_2/Si samples with various thickness, one SiN_x/Si sample, one $SiN_x/SiO_x/Si$ sample, one OPO (oxide/polysilicon/oxide) sample and one ONOPO (oxide/nitride/oxide/polysilicon/oxide) sample. All the results are summarized in Table 1. Experimental and simulated curves of the OPO sample measured with the SE300 in scanning mode are also reported in Fig. 3. As shown in the figure, the measurements extend up to 6.5 eV (190 nm) with a good accuracy. Results are also compatible to reference ES4G results. The coherence between scanning and IPDA mode is very good even if bottom SiO_2 thicknesses cannot be fitted in OPO and ONOPO structures.

3.2. Stability, reproducibility and repeatability of the system

These quantities have been intensively studied during the I300I gauge study [6]. The repeatability is evaluated repeating the measurement on one sample at one location without moving it. It is essentially

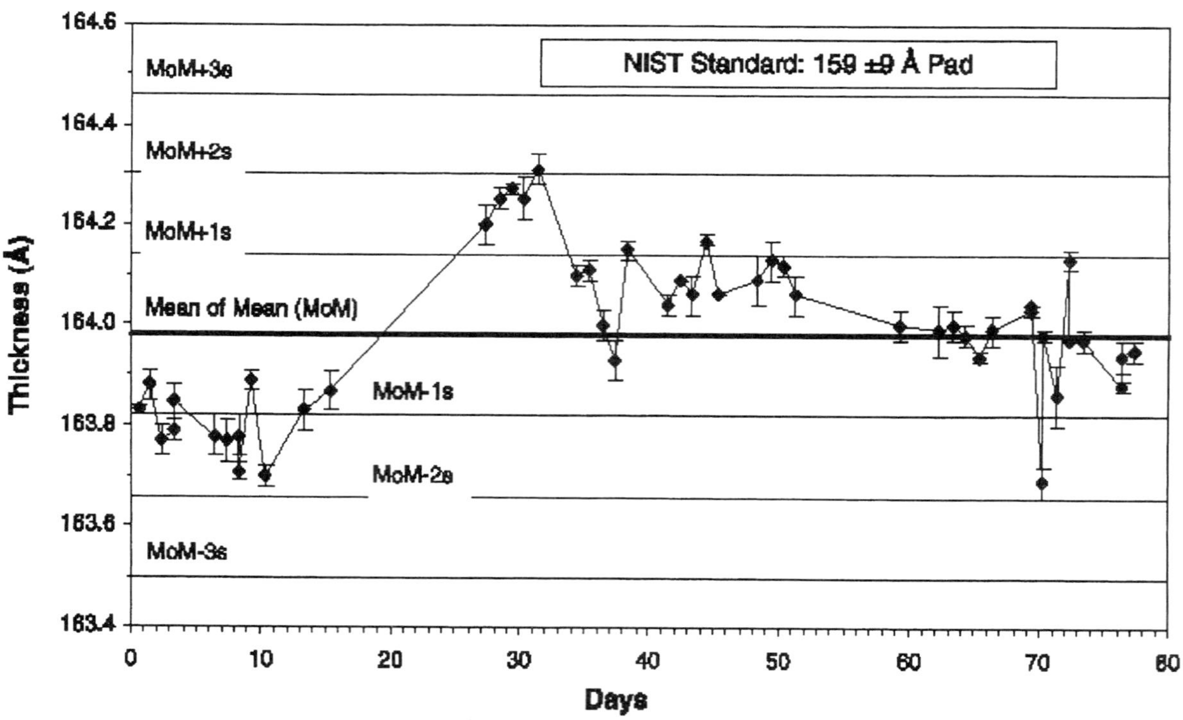

Fig. 4. Stability measurement on the 159±9 Å pad of a NIST wafer. Each data point shown is the average over three consecutive measurements. The error bars are calculated from minimum and maximum values of these three points.

dependent on the signal-to-noise of the measuring instrument. In our case, it is extremely good (only 0.02 nm for the $\pm 3\sigma$ deviation on a 200-nm thick SiO_2 layer on silicon). The reproducibility adds in the same procedure the unload/load of the wafer in between each measurement. So, the repeatability of all the other elements of the system (prealigner and X–Y–Z table) is tested. It is about ten times higher than the repeatability for the same kind of sample.

Finally, during the stability test, the same sample was measured in the same location three times continuously, every day during a long period. One example of such a measurement obtained on a NIST standard SiO_2/Si sample 159±9 Å thick is reported in Fig. 4. The measurement cover a period of 78 days interrupted only twice. The $\pm 3\sigma$ deviation is only 0.05 nm for this long period of measurement.

3.3. High-resolution, large-size mapping using SE300 system

One example of high-resolution mapping on a 300-mm implanted wafer is reported in Fig. 5. The measurement is made in IPDA mode on 49 points on the sample surface. The thickness of the perturbated zone is fitted assuming a homogeneous layer on top of monocrystalline silicon with optical indices extracted from a scanning measurement on the same sample.

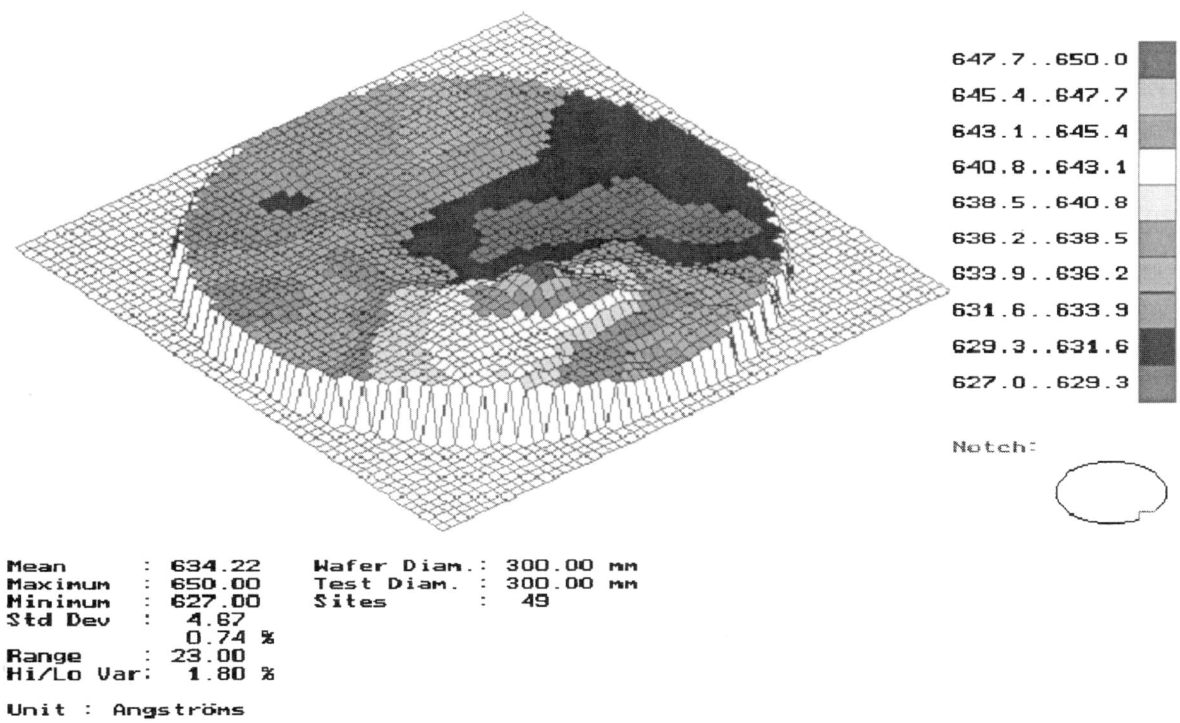

Fig. 5. High-resolution mapping obtained on a 300-mm implanted silicon wafer. The thickness of the perturbated zone is extracted point by point on the entire surface of the wafer.

4. Conclusion

A new tool for characterization of 300-mm wafers for the new generation of ICs has been presented. Based on a high accuracy spectrocopic ellipsometer, this instrument can characterize complex multilayers and thin films with an excellent accuracy. Its double spectrometer allows scanning measurements with very good wavelength resolution and also rapid multichannel measurements. This unique feature allows the end user to extract optical indices of unknown materials for subsequent analysis in IPDA mode. The wavelength range is extended in the deep UV down to 190 nm.

Future versions of the instrument will include additional analysis by infrared spectroscopic ellipsometry to get additional chemical information on the same samples. A version for flat panel display applications is also under development.

References

[1] F. Ferrieu, F. Bernoux, J.L. Stehle, Vide Couches Minces 233 (Suppl.) (1986) 17.
[2] D. Zahorski, J.L. Mariani, L. Escadafals, J. Gilles, Thin Solid Films 234 (1997) 412–415.
[3] P. Boher, J.L. Stehle, Mater. Sci. Eng. B37 (1996) 116.
[4] P. Boher, J.L. Stehle, MRS Fall Meeting, December 2–6, 1996, vol. 452.
[5] P. Boher, J.L. Stehle, EMRS Spring Meeting, Strasbourg, June 16–20, 1997.
[6] S. Wurm, SOPRA SE-300 Gauge Study Report, Technology Transfer 98033491-ENG, I300I, March 1998.
[7] T.R. Corle, SPIE 2439 (1995) 212–218.
[8] SOPRA patent.
[9] SEMI International Standards, Equipment Automation Software, Mountain View, CA, 1990.
[10] P. Durgapal, J.R. Ehrstein, N.V. Guyen, Paper 6, Session 4, International Conference on Characterization and Metrology for ULSI Technology, Gaithersburg, March 23–27, 1998.

ELSEVIER
Microelectronic Engineering 45 (1999) 277–282
MICROELECTRONIC ENGINEERING

Accurate infrared absorption measurement of interstitial and precipitated oxygen in p^+ silicon wafers

O. De Gryse[a,*], P. Clauws[a], L. Rossou[b], J. Van Landuyt[b], J. Vanhellemont[c]

[a]*University of Gent, Krijgslaan 281, B9000 Gent, Belgium*
[b]*University of Antwerpen-EMAT, Groenenborgerlaan 171, B 2020 Antwerpen, Belgium*
[c]*Wacker Siltronic AG, P.O. Box 1140, D-84479 Burghausen, Germany*

Abstract

A novel infrared absorption method has been developed to measure the interstitial oxygen concentration in highly doped silicon. Thin samples of the order of 10–30 μm are prepared in an essentially stress-free state without changing the state of the crystal. The oxygen concentration is then determined by measuring the height of the 1136-cm^{-1} absorption peak due to interstitial oxygen at 5.5 K. The obtained results on as-grown samples are compared with those from gas fusion analysis. The precipitated oxygen concentration in annealed samples is also determined with the new method. It will be shown that the interstitial oxygen concentration in highly doped silicon can be determined with high accuracy and down to concentrations of 10^{17} cm^{-3}.

1. Introduction

Heavily p-doped 300-mm Cz–Si wafers will be extensively used as substrates for epitaxial layers, in order to avoid problems arising from crystal defects in the near-surface region of Cz–Si. As the wafers of the 300 mm generation will be both sided polished, external gettering using the backside of the wafer becomes increasingly difficult, thus increasing the importance of internal gettering techniques. Although boron is an efficient getterer for different transition metals, it is expected that the additional gettering by SiO_x precipitates will be needed to increase yield. To study the behaviour of oxygen it is necessary to be able to determine the interstitial oxygen (O_i) concentration. For moderately doped Si-wafers Fourier transform infrared (FT-IR) spectroscopy is the standard technique. However, low resistivity material is not transparent in the mid-IR region because of the free carrier absorption which dominates the infrared light absorption. Other techniques, like gas fusion analysis (GFA) [1], have therefore been developed and are extensively used to determine the total oxygen content, leaving the oxygen configuration – precipitated or interstitial – uncertain. Another FT-IR technique has been developed to measure the O_i concentration in highly boron-doped silicon [2]. In this technique a highly boron-doped wafer is bonded on a thick high-resistivity float zone wafer

*Corresponding author.

0167-9317/99/$ – see front matter
PII: S0167-9317(99)00180-X

by a heat treatment at 1100°C in a nitrogen ambient and subsequently the highly doped part is then thinned and polished to a thickness for which the sample is transparent in the mid-IR region. The major disadvantage of this method is the need for a heat treatment at 1100°C, which can dissolve small pre-existing nuclei and/or lead to a growth of large preexisting nuclei or precipitates in an as-grown or annealed wafer, respectively. FT-IR techniques to measure the O_i concentration in highly antimony doped silicon at room temperature [3] down to 15 mΩ-cm and at low temperature down to 10 mΩ-cm [4] are known, too. The thinning and mirror polishing methods used by these authors limit the thickness to 160 and 65 μm, respectively. However, the required thickness for p-type silicon of the same resistivity is much lower.

In this contribution a novel method is outlined to measure the O_i concentration in heavily boron-doped silicon by preparing very thin specimens in an essentially stress-free state and by measuring the low temperature 1136 cm^{-1} absorption peak belonging to O_i [5].

2. Experimental

Czochralski-grown silicon samples doped with boron were used for this study as summarized in the upper part of Table 1. Oxygen concentration data for the highly doped as-grown wafers were available from GFA (middle part of Table 1) and will be compared with the results of the present investigation. The annealing sequence applied in the present investigation consisted of a first nucleation step (16 h at 500°C), a stabilization and growth step (16 h at 800°C) and a precipitation step (8 h at 1050°C). All anneals were performed in argon atmosphere. Prior to annealing the samples were etched (HNO_3:HF:CH_3COOH in a ratio of 5:1:1) in order to remove mechanical damage and contamination in the surface layer.

Table 1
Sample data and the interstitial and precipitated oxygen concentration as determined with FT-IR

	Sample					
	B1	B2	B3	B4	B5	Sb1
Dopant	Boron	Boron	Boron	Boron	Boron	Antimony
Resistivity [mΩ-cm]	18.8	18.6	17.8	8.7	2×10^4	26.4
As-grown						
Thickness (μm)	21.5	30.2	32	13.8	601	27.4
α(1136) (cm^{-1})	30.7±0.3	23.5±2.8	22.3±1	29±4	RT	28.4±0.9
$[O_i]$ (10^{17} cm^{-3})	7.1±0.6	5.5±0.8	5.2±0.3	6.7±1	8.09	6.6±0.6
GFA: $[O_i]$ (10^{17} cm^{-3})	8	5.5	–	6.9	–	6.87
Rel. dev.	0.11	0	–	0.03	–	0.04
Annealed						
Thickness (μm)	26.4	19.0	19.3	–	497	–
α(1136) (cm^{-1})	8.7±0.3	5.4±0.6	5.8±0.5	–	50.9	–
$[O_i]$ ($\times10^{17}$ cm^{-3})	2.0±0.1	1.3±0.2	1.3±0.1	–	6.23	–
$\Delta[O_i]$ ($\times10^{17}$ cm^{-3})	5.1±0.2	4.2±0.5	3.8±0.2	–	1.86	–

Infrared spectra were recorded with a BRUKER IFS-66v FT-IR instrument at 5.5 K and with a resolution of 1 cm^{-1}.

3. Interstitial oxygen determination in heavily doped silicon

As mentioned earlier a method using thin foils of highly doped silicon on a lowly doped supporting wafer has already been used before [2]. However, that procedure needs a high-temperature treatment which will change the thermal history and possibly also the O_i concentration of the sample. The alternative method presented here is the preparation of ultra-thin specimens in a stress-free state to avoid the need for a support or thermal treatment. The method is based on a procedure used in TEM-sample preparation with some adjustments to prepare samples with a larger area than usual for TEM. No temperature treatments are needed so that the material to be investigated remains unaffected.

The sample is mounted on a tripod polisher or a similar device [5] and is then thinned and mirror polished by making use of a series of abrasives with decreasing grain size. If both surfaces of the original sample are unpolished then one side needs to be polished first before thinning. Next, the up to 1-mm thick sample is put on a glass plate with the mirror-polished surface facing the glass plate. A drop of low-viscous cyanoacrylate adhesive (e.g. Loctite 420) is put on the glass plate and flows under the sample, leaving no air bubbles behind. This glue is very strong, can be dissolved in acetone, and the adhesive layer is very thin (3–6 μm). The absence of air bubbles between the glass plate and the specimen is of utmost importance. The glass plate with sample is then glued on a tripod polisher sample holder. The legs of the tripod are adjusted so as to position the glass plate parallel to the abrasive surface. Mechanical polishing induces stress in the surface layer of the sample, to a depth of up to approximately three times the abrasive grain size used. A series of abrasives with decreasing grain size is used in order to find a reasonable compromise between time needed for preparation and limiting the depth of the stress-induced layer. The polishing includes a pre-thinning step and a thinning and mirror-polishing step. In the first step the sample is thinned to 100 μm using a rough abrasive (grain size 20–30 μm until a thickness of 200 μm and grain size 6–12 μm until a thickness of 100 μm). In the second step the sample is thinned and mirror-polished to the desired thickness. The abrasive surface consists of 3 M diamond foils with successive grain sizes 6 μm (to within 25 μm of the desired thickness), 3 μm (to within 10–15 μm of the desired thickness), 1 μm (to within 10–15 μm of the desired thickness) and finally 0.5 μm. Samples prepared in this way are essentially stress free and have a thickness accurate to within a few μm of the desired final thickness. When finished, the sample holder with sample is put in an acetone bath to dissolve the glue. The released sample is then handled with a brush and conserved in a recipient containing isopropyl alcohol.

Prior to the infrared measurement the sample is placed with a wetted brush in a holder with a slightly larger area than the samples surface area, in order to take into account the different thermal expansion coefficient of silicon and of the material of the holder.

The IR spectra are recorded at 5.5 K. Interference fringes are visible in the high wavenumber region of the transmission spectrum. The average spacing between the fringes is used to calculate the thickness of the sample [6]:

$$d = \frac{1}{2 \cdot n \cdot \langle \Delta\sigma \rangle} \quad (1)$$

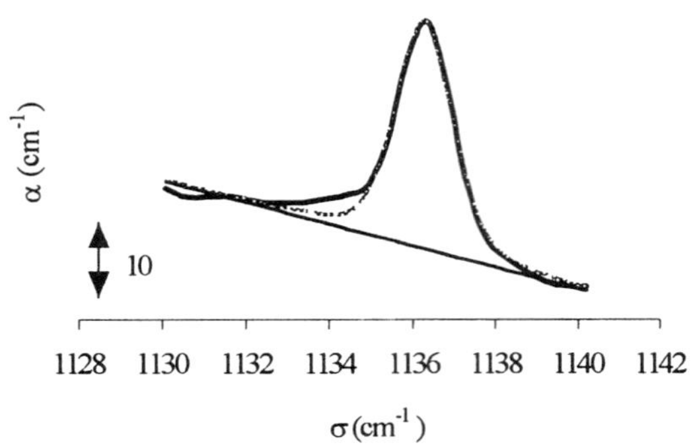

Fig. 1. Absorption spectrum of sample B1 (full curve) together with the fitted lineshape function (dashed curve) and accompanying baseline (straight line).

with d the thickness of the sample (cm), n the index of refraction of silicon (3.417 at RT) and $\Delta\sigma$ the wavenumber spacing between the interference fringes (cm^{-1}). On the other hand, the interference fringes are not visible in the wavenumber region 1130–1140 cm^{-1}, where the free carrier absorption is very strong (see Fig. 1). The absorption belonging to O_i can be approximated with good accuracy by:

$$\alpha_{O_i} = \frac{1}{d} \cdot \ln\left(\frac{1}{T}\right) - \text{baseline} \tag{2}$$

The height of the 1136.3 cm^{-1} peak is determined by fitting a pre-defined lineshape function and a baseline to the experimental spectrum in the frequency region 1130–1140 cm^{-1} (Fig. 1). The O_i concentration is calculated as follows: $[O_i] = F_{RT} \times H \times \alpha$ (1136.4 cm^{-1}) [cm^{-3}], with $H = \alpha$ ($\sigma =$ 1107 cm^{-1}; $T =$ RT)/α ($\sigma =$ 1136.3 cm^{-1}; $T = 5.5$ K) the conversion factor for the 5.5 K absorption to the RT standard, and F_{RT} ($= 3.14 \times 10^{17}$ cm^{-2} [7]) the IOC88 standard calibration conversion factor for RT. For a resolution 1 cm^{-1} before apodisation and using the Blackman–Harris three-term apodisation function the value of H is equal to $H = 0.0707 \pm 0.0033$. The determination of H for several resolutions and apodisation functions will be published elsewhere [8].

4. Comments on the thickness of the sample

Two conflicting trends have to be taken into account when estimating the optimal thickness of the sample. The minimum thickness d_{min} is determined by the detection limit for the 1136.3 cm^{-1} absorption peak, which in the actual case was found to be given by the condition: α(1136.3 cm^{-1})$\times d \gtrsim 0.01$. On the other hand, an acceptable signal-to-noise ratio requires a sufficiently thin and transparent sample; a minimum transmission at 1136 cm^{-1} of 0.025 is considered as a practical limit. An estimate of the maximum thickness d_{max} allowed to meet this requirement may be calculated from the free carrier infrared response according to the Drude model [6], given the sample resistivity. Material parameters have been taken from the literature and from fittings to the experimental transmission spectra in the present study. The procedure allows to estimate the maximum thickness for a given resistivity. More details will be published elsewhere [9].

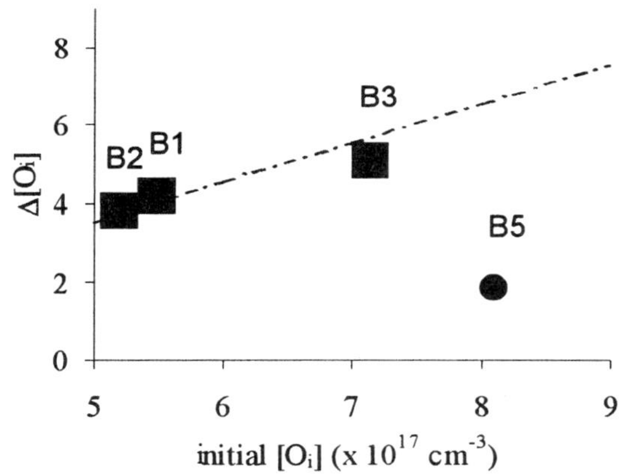

Fig. 2. Diagram of $\Delta[O_i]$ vs. initial $[O_i]$. The dashed line is the maximum amount of O_i which can be precipitated (=initial $[O_i]-1.45\times10^{17}$ cm^{-3}). After the thermal anneal nearly all available oxygen is precipitated for the highly doped samples (■) in contrast to the moderately doped sample (●).

5. Results and discussion

The O_i concentrations for the as-grown wafers obtained with the FT-IR on the thin specimens are shown in middle part of Table 1, together with the GFA data and the relative deviations with the FT-IR measurements. The FT-IR results agree very well with the total oxygen concentration obtained with GFA. The uncertainty intervals for the absorption coefficients are determined by calculating the root mean square of the noise in the region 1120–1150 cm^{-1}. Although the result of the 8.7 mΩ-cm sample (B4) shows a larger uncertainty interval, due to the higher noise level in the absorption spectrum, it still agrees very well with GFA data. It is clear that even for wafers in the 10 mΩ-cm region FT-IR can still be used to determine the O_i concentration with an accuracy of $\pm1\times10^{17}$ cm^{-3}.

The O_i concentration of the annealed samples is also determined with good accuracy (lower part of Table 1). The data show once more that for resistivities down to 18 mΩ-cm, the method can be used to determine $[O_i]$ in highly boron-doped silicon samples down to concentrations of about 1×10^{17} cm^{-3}.

For the precipitation growth step at 1050°C the solubility concentration for O_i is 1.45×10^{17} cm^{-3}. The dashed line in Fig. 2 represents the difference between the initial $[O_i]$ and the solubility limit, i.e. the maximum) $\Delta[O_i]$. The oxygen precipitation process in the annealed, 18 mΩ-cm p^+-type silicon wafers is close to saturation (■ in Fig. 2). As compared with the precipitated oxygen concentration of the moderately doped wafer (● in Fig. 2) the precipitation process in highly boron-doped silicon is substantially more efficient, even in the case of a lower initial $[O_i]$. The latter is in agreement with previous observations [10,11].

6. Conclusions

A new method has been developed to determine the O_i in p^+ silicon material by measuring the height of the low temperature 1136 cm^{-1} absorption peak in very thin samples. The method has the advantage that the state of the sample is not altered by the preparation procedure. The good agreement between the O_i concentrations in as-grown material determined with the new FT-IR technique and GFA data, shows that the procedure yields reliable results. Interstitial oxygen concentrations down to approximately 10^{17} cm^{-3} can be measured with sufficient accuracy in materials with resistivities down to 8 mΩ-cm. This opens the possibility to study the precipitation process in heavily doped material by measuring changes in O_i concentration after thermal treatment.

Acknowledgements

This work was performed with the financial support of the National Science Foundation (FWO) (project number: FWO-G.0051.97).

References

[1] H. Walitzki, H.-J. Rath, J. Reffle, S. Pahlke, M. Blätte, Electrochem. Soc. Proc. 86–84 (1986) 86.
[2] M. Koizuka, M. Inaba, H. Yamada-Kaneta, Mater. Res. Soc. Symp. Proc. 442 (1997) 31.
[3] A. Borghesi, M. Geddo, G. Guizzetti, P. Geranzani, J. Appl. Phys. 68 (1990) 1655.
[4] Q.Y. Wang, T.H. Cai, Y.H. Yu, L.Y. Lin, Semicond. Sci. Technol. 12 (1997) 464.
[5] J.P. Benedict, S.J. Klepeis, W.G. Vandygrift, Ron Anderson, EMSA Bull. 19 (1989) 74.
[6] C.F. Bohren, D.R. Huffman, in: Absorption and Scattering of Light by Small Particles, Wiley, New York, 1983.
[7] A. Baghdadi, W. Bullis, M. Croarkin, Y.-Z. Li, R. Scace, R. Series, P. Stallhofer, M. Watanabe, J. Electrochem. Soc. 136 (1989) 2015.
[8] O. De Gryse, P. Clauws, to be submitted to J. Electrochem. Soc.
[9] O. De Gryse, P. Clauws, L. Rossou, J. Van Landuyt, J. Vanhellemont, W. Mondelaers, J. Electrochem. Soc., submitted to Rev. Sci. Instrum.
[10] S.K. Bains, D.P. Griffiths, J.G. Wilkes, R.W. Series, K.G. Barraclough, J. Electrochem. Soc. 137 (1990) 647.
[11] T. Ono, E. Asayama, H. Horie, M. Hourai, M. Sano, H. Tsuya, K. Nakai, Jpn. J. Appl. Phys. Part II 36 (1997) L249.

ELSEVIER

Microelectronic Engineering 45 (1999) 283–289

MICROELECTRONIC ENGINEERING

In-line monitoring of 300 mm silicon epitaxial and CZ wafers using surface charge profiler

E. Kamieniecki[a,*], T. Bickl[b], J. Tower[a]

[a]QC Solutions, Inc., Woburn, MA 01801, USA

[b]EURIS GmbH, Marktplatz 2-4, D-61118 Bad Vilbel, Germany

Abstract

The surface charge profiler (SCP) offering non-contact electrical characterization of the near-surface region of silicon wafers is discussed. The system permits fully automatic handling of 300- and 200-mm wafers. The SCP method, based on a low intensity illumination a.c. surface photo-voltage principle, does not require any surface preparation. It allows for a fast (600 points/min), high-resolution mapping of the active doping concentration in the near-surface region as well as surface recombination lifetime. The capabilities of the SCP method for process monitoring and development are illustrated with 200- and 300-mm wafers, focusing on the effects of epi growth conditions on the layer uniformity and its resistivity.

1. Introduction

Effective in-line, real-time process monitoring is required to ensure adequate process reproducibility as well as early detection of potential process malfunction. At the same time, process monitoring is limited by the high cost of sacrificial monitor wafers. This is becoming especially critical with emerging use of 300-mm wafers. In particular, there is a critical need for a reliable non-contact method of monitoring epitaxial layer resistivity in the production of 300-mm wafers as well as the near-surface contamination and damage in silicon wafer processing. To fill the gap of non-contact resistivity measurement, several techniques have been proposed recently, based on air gap MIS C–V [1], air gap MIS $C_{a.c.\text{-}SPV}$–V [2], and Air Gap a.c.-SPV [3] (surface charge profiler (SCP) method discussed in this paper). Table 1 summarizes the main characteristics of each method for production monitoring of epitaxial wafers. As a comparison, the conventional Schottky Barrier MS C–V technique is reported. New techniques developed for in-line, non-contact measurements of resistivity must be at least as accurate as existing methods (mercury probe, four-point probe, spreading resistance) and free of their weaknesses: wafer damage, slow measurement speed, and poor

*Corresponding author. Tel.: +1-781-938-4030; fax: +1-781-933-0675.
E-mail address: emil@qcsolutions.com (E. Kamieniecki)

0167-9317/99/$ – see front matter
PII: S0167-9317(99)00188-4

Table 1
Resistivity of epi layers, comparison of measurement techniques

Technique	Schottky barrier MS C–V	Air gap MIS C–V	Air bearing MIS $C_{a.c.\text{-}SPV}$–V	SCP: air gap a.c.-SPV
Contact	Yes	No	No	No
Probe-to-wafer distance	0	0.5 μm	0.5 μm	100 μm
High electric field across air gap	–	Yes	Yes	No
Wafer damage: Contamination, susceptible electric discharge, ionic and charged particles	Yes	Yes	Yes	No
Measurement time	10–30 min including surface preparation	6 min per point	7 min per five points	600 points per min
Fast mapping	No	No	No	Yes
Doping profile	Yes	Yes	Yes	No
Measure near-surface doping	No	No	No	Yes
Application	Diagnostic	Diagnostic	Diagnostic	In-line

repeatability. We present here the SCP method with these objectives, as a new tool for fast non-contact doping mapping of epitaxial and bulk wafers. We shall discuss main advantages and limitations of this technique for high-purity silicon control and present experimental data obtained on SCP 7300 and 7200 systems manufactured by QC Solutions.

2. Principle of the SCP method

We will only briefly summarize the operation principles of the SCP method. For a detailed description refer to [3]. If the semiconductor surface is in an inversion state, the depletion layer width, W, reaches its maximum value, W_{max}, which is related to the doping concentration in the near-surface region via Eq. (1) [4]:

$$W_{max} = \sqrt{\frac{4\varepsilon_{Si}\, kT \ln(\mathrm{Nsc}/n_i)}{q^2 \mathrm{Nsc}}}, \tag{1}$$

where q is the elementary charge, k is Boltzmann's constant, T is absolute temperature, ε_{Si} is the dielectric constant of silicon, n_i is the intrinsic carrier concentration in the silicon at temperature T, and Nsc is the net average doping in the depleted layer. In the SCP method, the depletion layer width, W, is determined by measurement of the surface photo-voltage induced by a low intensity modulated light: a.c.-SPV signal [3,5]. From the analysis of the real and imaginary part of the a.c.-SPV signal,

δV_S, the surface recombination lifetime, τ_S, and the depletion layer width, W, are simultaneously derived:

$$\frac{1}{\tau_S} = -\omega \frac{\text{Re}(\delta V_S)}{\text{Im}(\delta V_S)} \tag{2}$$

$$W = -\left(1 + \frac{1}{\omega^2 \tau_S^2}\right) \frac{\omega \varepsilon_{Si}}{q\Phi} \text{Im}(\delta V_S) \tag{3}$$

where ω is an angular frequency of light intensity modulation and Φ is the absorbed photon flux. From W_{max} measurement, Nsc is calculated using Eq. (1) and then converted to resistivity using ASTM standard formulas.

3. Description of the instrument

The principle described above is implemented in a commercial instrument developed by QC Solutions, and introduced in 1994 for in-line monitoring of cleaning processes [6] and doping measurements [7–10]. Advantages of this technique, especially for doping measurements, are discussed and illustrated with results obtained on the SCP systems. An overall view of the 300-mm system (Model 7300) is shown in Fig. 1. The functional diagram of the SCP system is shown in Fig. 2.

During the measurement cycle the wafer is positioned on a chuck that moves underneath the probe head. Low-intensity incident light originates from a blue LED and shines through the transparent sensor onto the wafer surface. The a.c.-SPV signal is picked up by the sensor, amplified in the probe and further processed in the data acquisition system. The distance of the sensor from the wafer (air gap) is controlled through a feedback system. This real-time control of the air gap minimizes sensitivity to wafer and stage flatness and to the mechanical vibrations during wafer mapping.

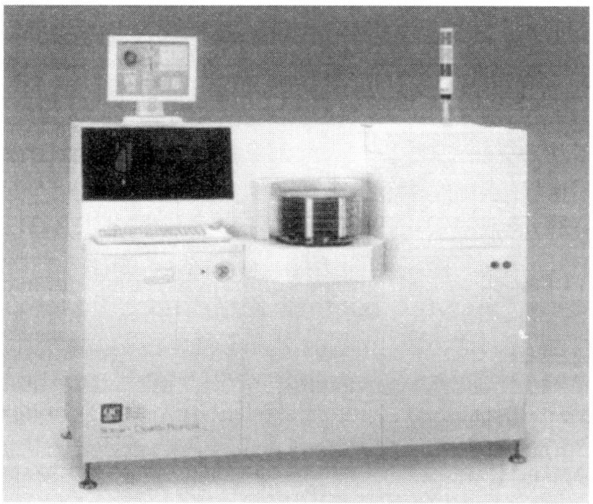

Fig. 1. QC solutions SCP 7300 capable of handling 200- and 300-mm wafers.

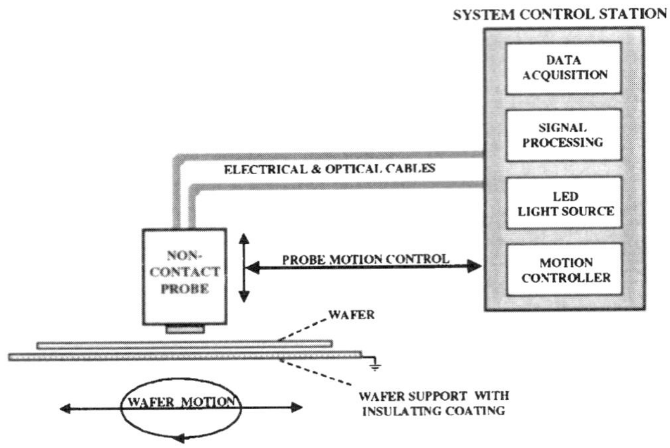

Fig. 2. Functional diagram of the SCP system.

4. Results

4.1. High speed non-contact mapping

Wafers are measured under very low optical flux without any external bias voltage or corona charging. Therefore, in addition to being non-invasive, the SCP is much faster than other techniques. This enables the SCP to map the doping and surface recombination lifetime of the entire wafer in a few minutes, depending on the wafer size and measurement resolution. As an example, the maps in Fig. 3 show highly non-uniform doping regions of epitaxial wafers. The map on the left of Fig. 3

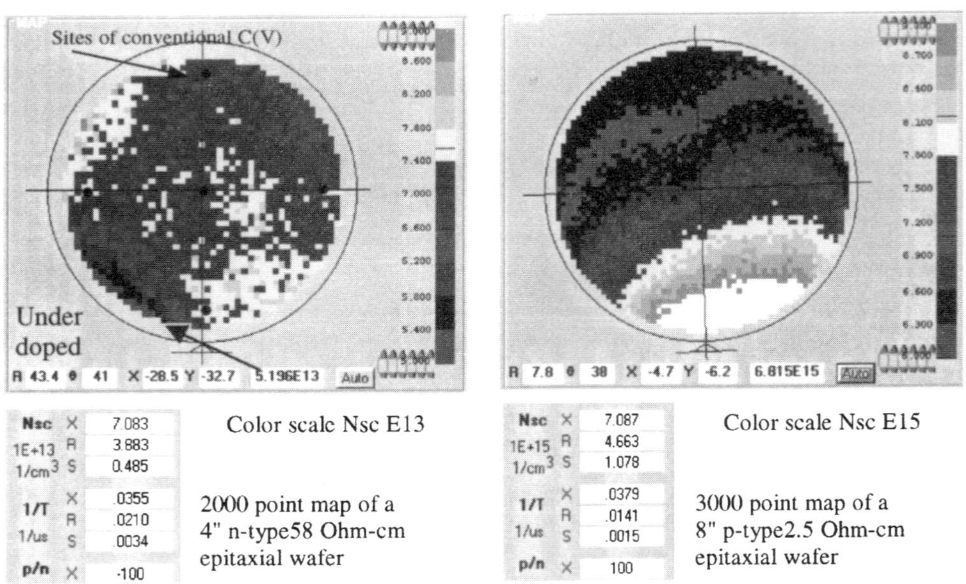

Fig. 3. Doping map of non-uniform epitaxial wafers.

shows an under-doped region that would not be detected by routine five-point destructive C–V analysis. The map on the right of Fig. 3 shows detection of non-uniform doping in an epitaxial wafer processed without susceptor rotation [11,12]. Another example of doping in epitaxial wafers is shown in Fig. 4, for the case of 300-mm wafers. In this example, the effects of a backseal during epitaxial growth are shown for p-type layers grown on p^+ substrates. The use of backseal prevents transfer of dopant from the substrate to the epi layer and results in more uniform doping.

4.2. Sensitivity to near-surface doping

In the case of good quality epitaxial layers, the relationship between SCP and MS C–V measurements is linear for doping concentrations from about 10^{13} to 10^{18} at/cm^{-3} (0.1–1000 Ω-cm), with the correlation factor R^2 exceeding 0.99. However, the SCP measured doping may differ from conventional measurements (C–V, four-point probe) when the doping in the subsurface region (within a few Debye lengths from the surface) differs from the doping in the bulk of the wafer. Unlike C–V methods that measure carrier concentration at the edge of the depletion layer when the width of the layer exceeds 3–5 Debye lengths [4] (about 1 μm at doping of 10^{15}/cm^3), the SCP method measures the average doping concentration in the entire depletion layer, including the subsurface region (see Fig. 5). This sensitivity to the subsurface can be used to monitor surface overdoping that might occur at the end of the epitaxial growth process (see Fig. 6), as well as deactivation of boron due to pairing or compensation during cleaning steps and thermal processes. In some cases, this deactivation can also be related to surface contamination [10].

The sensitivity of the SCP to the surface overdoping has been confirmed in a number of evaluations using epitaxial wafers of different sources. Fig. 6 shows an example of overdoping in four epitaxial wafers where the growth conditions at the end of the runs were intentionally changed. These growth process changes resulted in substantial differences in the subsurface doping concentration. The subsurface region of the epitaxial wafer is the most critical for state of the art-integrated circuits. Conventional C–V testing did not detect these critical differences in subsurface doping.

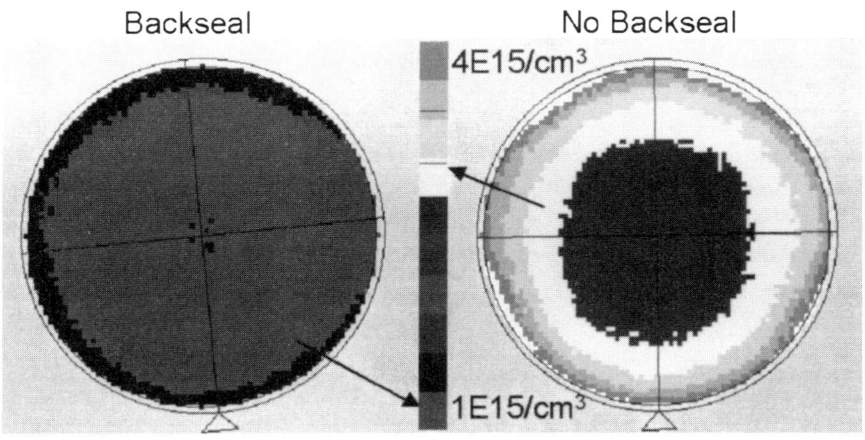

Fig. 4. Doping concentration maps of 300-mm p-type epitaxial wafers, showing the effects of backseal on dopant incorporation.

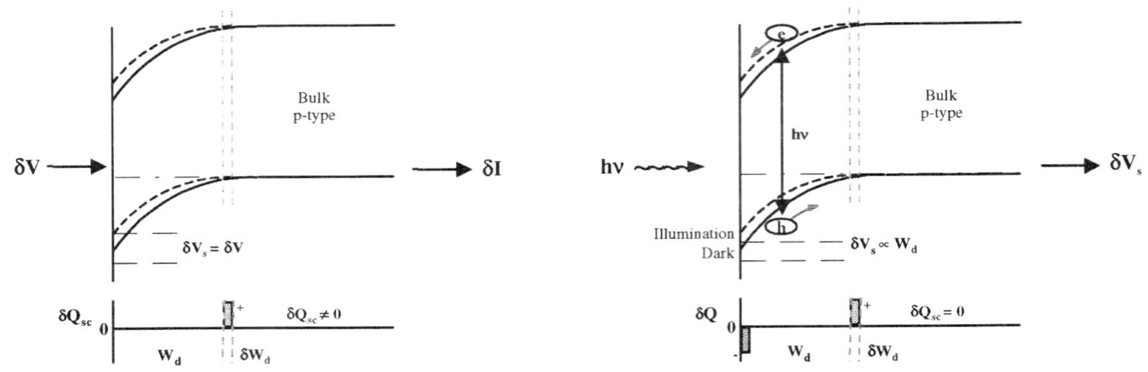

Fig. 5. Comparison of band diagrams for C–V and SCP measurements.

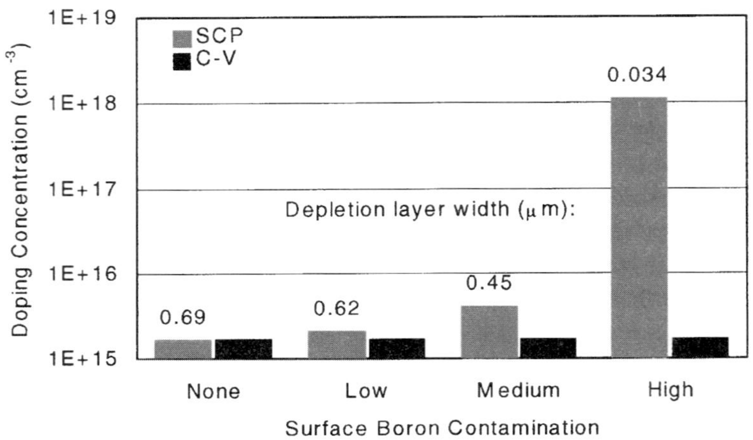

Fig. 6. Detection of subsurface overdoping using SCP. C–V was unable to detect differences between these wafers.

5. Conclusion

The non-contact SCP method has been used for characterization of unpatterned production wafers, epitaxial wafers in particular. Advantages of this technique are discussed in detail for doping measurements of epitaxial layers. The main advantages pointed out were: (1) non-contact measurement, which reduces need for destructive testing in production epitaxial wafers; (2) fast mapping capability, which makes SCP an ideal in-line monitoring tool; (3) sensitivity to near-surface doping, which enables a tighter epi-process control.

References

[1] DaiNippon Screen: Non-Contact C–V Measurement System CV-3000 Technical Information, 1997.
[2] S. Liberman, SPIE 3215 (1997) 35.

[3] E. Kamieniecki, G.J. Foggiato, in: W. Kern (Ed.), Handbook of Semiconductor Wafer Cleaning Technology, Noyes Publications, 1993, pp. 497–536.
[4] S.M. Sze, in: Physics of Semiconductor Devices, Wiley, New York, 1981, ch. 7.
[5] E. Kamieniecki, J. Vac. Sci. Technol. 20 (1982) 811–814.
[6] E. Kamieniecki, P. Roman, D. Hwang, J. Ruzyllo, in: M. Heyns, M. Meuris, P. Mertens (Eds.), Second Intern. Symp. on Ultra-Clean Processing of Silicon Surfaces, Leuven, Belgium, 1994.
[7] J. Ruzyllo, P. Roman, J. Staffa, I. Kashkoush, E. Kamieniecki, in: A. Iturralde, T. Lin (Eds.), The SPIE Proceedings Series, Process, Equipment, and Materials Control in IC Manufacturing II, Vol. 2876, SPIE, Bellingham, WA, 1996, p. 162.
[8] A. Danel, T. Lardin, G. Kamarinos, F. Tardif, in: B.O. Kolbesen, P. Stallhofer, C. Claeys, F. Tardif (Eds.), Second Intern. Symp. On Crystalline Defects and Contamination, The Electrochemistry Society Proceedings Series, Pennington, NJ, PV 97-22, 1997, p. 394.
[9] P. Roman, J. Staffa, S. Fakhouri, M. Brubaker, J. Ruzyllo, K. Torek, E. Kamieniecki, J. Appl. Phys. 83 (1998) 2297–2300.
[10] A. Danel, F. Tardif, G. Kamarinos, in: E-MRS Meeting, Strasbourg, France, Surface Dopant Concentration Measurement Using the Surface Charge Profiler Method: Characterization of Hydrogen and Metallic Contamination in Silicon, June 1998.
[11] A. Danel, F. Tardif, G. Kamarinos, M.C. Nguyen, Use of Non-Contact Resistivity Measurements for Epitaxy: Surface Charge Profiler Method, ECS Fall Meeting, Boston, MA, Nov. 1998 (in press).
[12] M.C. Nguyen, A. Danel, Use of Non-Contact Techniques in Characterization of Silicon Epi Growth Process, ECS Fall Meeting, Boston, MA, Nov. 1998 (in press).

ELSEVIER

Microelectronic Engineering 45 (1999) 291–297

MICROELECTRONIC ENGINEERING

Deep submicron 3D surface metrology for 300 mm wafer characterization using UV coherence microscopy

Paul C. Montgomery[a,*], D. Montaner[b]

[a]*Laboratoire PHASE, CNRS, 23, rue du Loess, BP 20, 67037 Strasbourg Cédex 02, France*
[b]*LEPSI, IN2P3/CNRS-ULP, 23, rue du Loess, BP 20, 67037 Strasbourg Cédex 02, France*

Abstract

With the increase in diameter of Si wafers to 300 mm and above, and the decrease in minimum feature size to the deep submicron, there is a need to improve the lateral resolution of optical inspection techniques. Coherence microscopy normally uses a white light source in order to produce a narrow fringe envelope for use in surface relief measurement. In this paper we show that high intensity coherent UV light sources can also be used if a large numerical aperture objective is employed, in order to improve the lateral resolution. This forms the basis for a new metrology tool that could find many applications in the inspection of large Si wafers.

Keywords: Coherence microscopy; White light interferometry; Non-destructive testing; Defect characterisation; Critical dimensions

1. Introduction

Optical lithography is now expected to be capable of being used up to and perhaps beyond 100 nm technology in semiconductor microelectronics fabrication. The continuing reduction in minimum geometric feature sizes has been made possible due to a number of technical advances in optical lithography. For example there have been improvements in exposure tools using wavelength reductions and off-axis computer controlled illumination, and in mask techniques using phase shift masks and optical proximity correction. The present 250 nm technology makes use of i-line illumination at 365 nm, and the development of the next generation 180 nm technology is based on DUV (deep UV) illumination at 248 nm using the KrF laser. The transition to 150 nm, expected in 2001 [1] is hoped to be achieved by using 193 nm (ArF laser) exposure tools on several mask layers.

With the move towards larger diameter wafers to 300 mm and above, and smaller feature sizes, the overall increase in the number and complexity of processing steps means that there will be greater problems resulting from wafer misprocessing, particularly if the older style of fixed process 'recipes'

*Corresponding author. Tel.: +33-3-88-106-231; fax: +33-3-88-106-293.
E-mail address: monty@phase.c-strasbourg.fr (P.C. Montgomery)

0167-9317/99/$ – see front matter
PII: S0167-9317(99)00194-X

with open loop control and offline metrology are retained. There is a growing need for new in situ sensors to provide real time data for use in active process control and wafer inspection ('sensor-driven model-based integrated manufacturing').

While the optical lithographic process itself has benefited greatly from various resolution enhancement techniques to produce smaller features, similar improvements in lateral resolution in optical inspection techniques used at the industrial level have not been able to follow at the same pace. Although optical microscopy has been used for some time for checking critical dimensions to within a few nm, it is only just recently that commercial reflection microscopes have been introduced that make use of UV illumination, quartz optics and UV sensitive cameras to provide lateral resolutions as small as 0.1 μm.

Coherence microscopy (or vertical scanning interference microscopy) is a fairly recent technique that has been used successfully for checking critical dimensions and misregistration measurements in automated wafer inspection systems [2] and for high throughput surface shape measurement [3,4] for process control. Using broadband, low coherence light in an interference microscope, a narrow envelope of fringes is scanned over the surface relief, giving nanometric depth discrimination and an improved lateral resolution compared with conventional microscopy.

In order to be able to resolve smaller features shorter wavelength sources need to be used but these tend to have a longer coherence length, which widens the fringe envelope and decreases the axial resolution. While broadband light is usually used to produce the narrow envelope of fringes, we have recently shown using red laser light [5] that the shape of the fringe envelope can also be controlled using the short depth of field associated with high numerical aperture objectives. High intensity coherent UV sources can therefore be used in coherence microscopy to improve the lateral resolution, as in optical lithography. In this paper we present some theoretical results for the use of UV light in coherence microscopy as a basis for a new generation of metrology tools for inspection of large 300 mm wafers.

2. Theory

2.1. The optical 'coherence probe'

Experimental details of coherence microscopy have been given elsewhere, the system consisting of a metallographic reflection microscope with interference objectives of the Michelson, Mirau or Linnik type [2–4,6]. An image of the surface of the sample with interference fringes superimposed is digitised using a CCD camera and imaging board. The sample is scanned axially in relation to the reference mirror using a piezo-electric positioner to control either the position of the reference mirror, the whole objective or the sample itself. A position sensor of the strain gauge, LVDT (linear variable differential transformer) or capacitive type is used in closed loop control to provide linear movement.

The narrow envelope of fringes (Fig. 2) is then used as an optical probe by identifying the point of maximum coherence between the light in the reference beam and that of the object beam at the peak of the envelope using image processing. Because the position of the probe plane is known from the calibrated position sensor, the position of the surface at that point can be determined.

The axial resolution of the coherence probe depends on the sharpness of the fringe envelope and therefore on the envelope width which is determined either by the coherence length of the light source or the depth of field of the objective, whichever is the shorter. The depth of field can be determined by considering the three dimensional intensity distribution near the image plane due to a point source in the object plane [7]. The intensity distribution along the optical axis Z of a microscope in the image plane, $I'(z')$, normalised to unity for $z'=0$ is described by:

$$I'(Z) = [(\sin Z)/Z]^2$$

where $Z = (\pi A'^2 z')/(2\lambda n')$ and A' and n' are the numerical aperture and refractive index in the image space. The depth of the image is generally accepted to be when the intensity falls to 80% of the maximum value, which occurs when $z' = \pi/4$. The associated depth of field d_f of the objective near the object plane can be determined to be:

$$d_f = \frac{(n\lambda)}{A^2}$$

where n is the refractive index of the medium between the lens and the object and A is the numerical aperture given by $A = n \sin \sigma$, and σ is the object-side aperture angle.

Fig. 1 shows the variation in depth of field of the objective, and therefore the fringe envelope width at the 80% intensity value, with wavelength down to a minimum wavelength of 100 nm, for different numerical apertures. The typical wavelengths used in optical lithography are indicated for comparison.

2.2. Controlling the width of the 'coherence probe'

Up until now, the coherence of the light 'probe' used in coherence microscopy has been limited by using a broadband source (typically a halogen or xenon arc lamp), giving a coherence length smaller than a micron. The advantage of this configuration is that a narrow envelope is retained whatever the numerical aperture, the envelope size being largely determined by the coherence length of the source

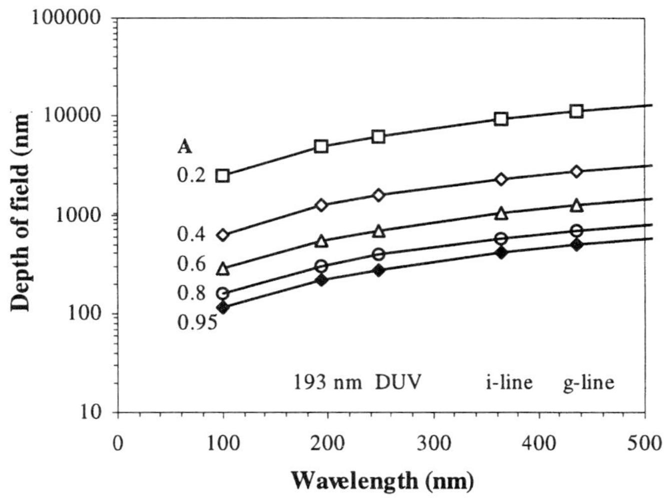

Fig. 1. Variation in theoretical depth of field, d_f, with wavelength for different numerical apertures for coherence microscopy.

which is shorter than the depth of field of the objective. Fig. 2 illustrates this effect with some simulated results of typical white light fringes (wavelength range of 0.4 μm to 1.1 μm) having a coherence length of 0.63 μm. The fringe envelope can be observed to be narrower than both the depth of field of 12 μm for the small numerical aperture of 0.25 (Fig. 2a) and the depth of field of 0.83 μm for the large numerical aperture of 0.95 (Fig. 2b).

But for high magnification work, the short depth of field associated with large numerical apertures can also be used to limit the coherence of the light instead of the spectral bandwidth of the source, illustrated in Fig. 3 with the use of 248 nm wavelength monochromatic light. In this case the coherence length, which is greater than 10 cm, is much greater than the depth of field, so that it is the depth of field function that determines the fringe envelope width, giving a very narrow fringe envelope (Fig. 3b).

The inconvenience of this arrangement is that for lower power objectives used to give a larger field of view (for example ×5 to ×20 objectives) the fringe envelope is wider (Fig. 3a), which decreases the axial resolution. But for high magnification work, in which large numerical apertures are used

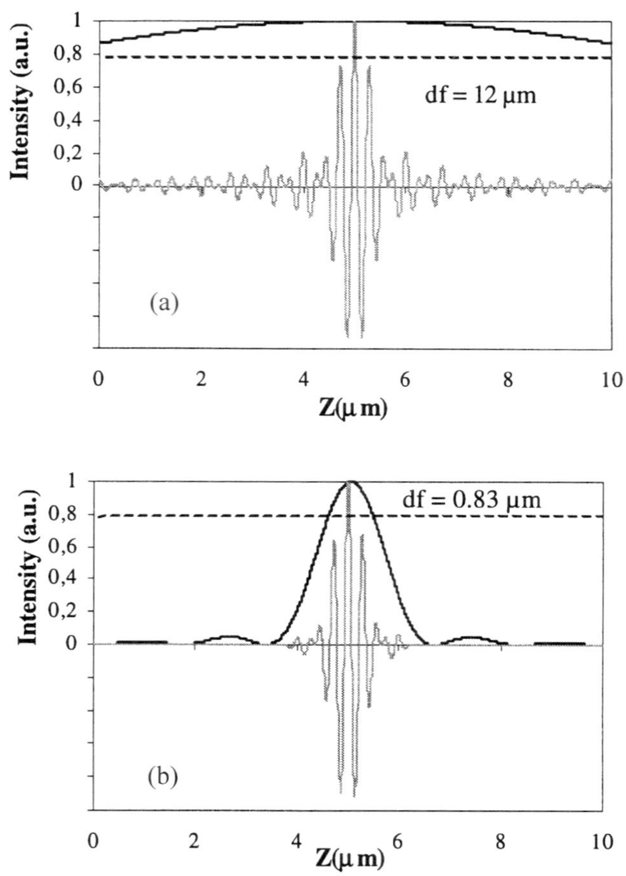

Fig. 2. Comparison of theoretical depth of field functions and axial intensity scans of fringes for white light coherence microscopy (λ_{av} = 750 nm). (a) ×10 objective (A = 0.25); (b) ×100 objective (A = 0.95).

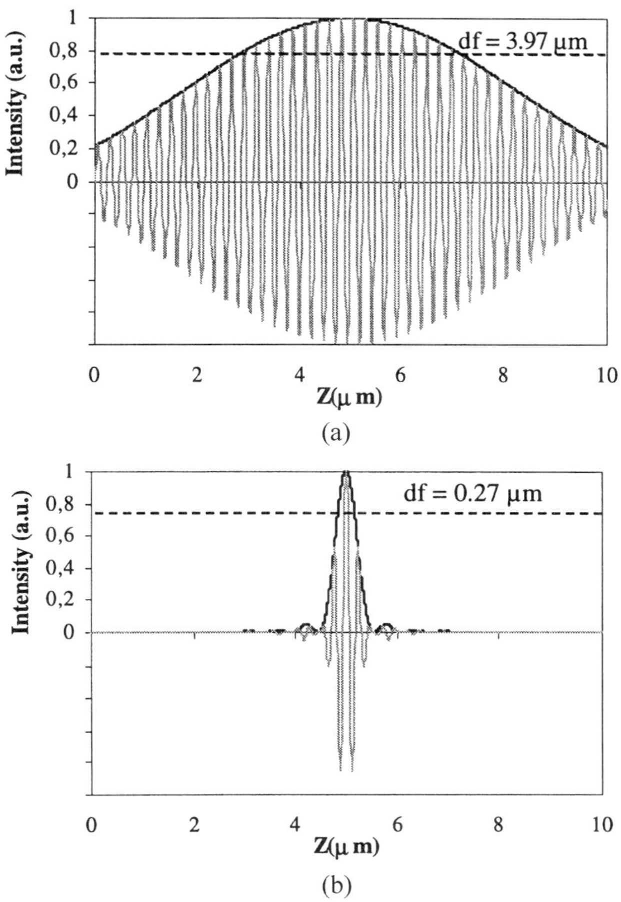

Fig. 3. Comparison of theoretical depth of field functions and axial intensity scans of fringes for DUV coherence microscopy (λ=248 nm). (a) ×10 objective (A=0.25); (b) ×100 objective (A=0.95).

(typically A=0.8 to 0.95) this opens the way to the use of much shorter wavelength sources which tend to have shorter coherence lengths, in order to improve the lateral resolution. For coherent light sources such as lasers, in practice, some additional coherence reduction is required together with the short depth of field to limit the spatial coherence [5].

2.3. Lateral super-resolution

Coherence microscopy has been demonstrated to show lateral super-resolution [8], the very high depth discrimination improving the lateral resolution for certain surface shapes such as planar structures. Davidson et al. [2] claim that the lateral resolution is similar to that of confocal microscopy, improved by a factor of about 1.2 compared with ordinary microscopy, the point spread function being the square of that of ordinary microscopy. They also show an example of intensity scans of a cleaved edge showing an improvement in resolution of over a factor of two. Lee and Strand [9] claim that coherence microscopy can provide twice the lateral resolution of conventional

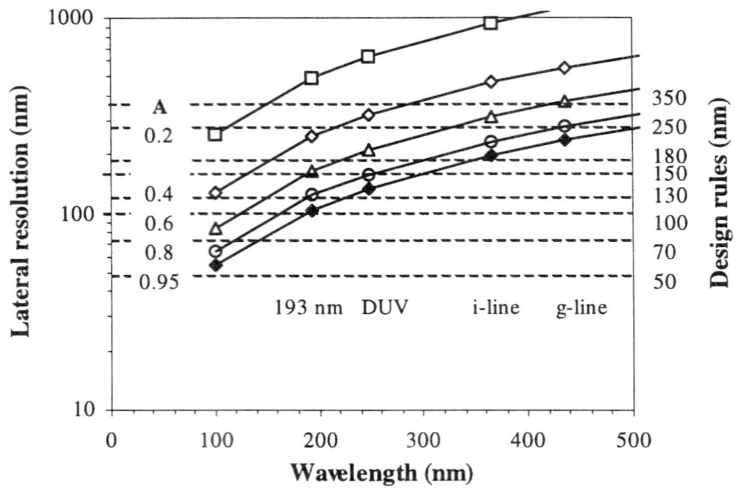

Fig. 4. Variation in theoretical lateral resolution with wavelength for different numerical apertures for coherence microscopy.

microscopy. If we retain the former more reserved figure of 1.2, then the lateral resolution, d_r of a coherence microscope is $d_r = (0.51\lambda)/nA$.

The variation of lateral super-resolution with wavelength for different numerical apertures is given in Fig. 4. These results illustrate that for a maximum numerical aperture of 0.95 under dry conditions (i.e. not using an immersion objective), wavelength reduction can be used to attain lateral resolutions below 100 nm, assuming the availability of the appropriate illumination and imaging optics and UV sensitive detector.

3. Potential applications of UV coherence microscopy in wafer inspection

Coherence microscopy is particularly well adapted to the dimensional metrology of deep planar features of microelectronic structures. With improvements in the lateral resolution by means of wavelength reduction, which appear to be possible to below 100 nm, UV coherence microscopy could therefore represent the basis for a new generation of optical inspection tools. It could certainly be used in offline characterisation, and because of its particular flexibility and parallel data acquisition capability has potential to be adapted to in situ measurement as well.

Because of the nanometric depth discrimination of coherence microscopy, the technique is already used in offline automated critical dimension measurements of semiconductor lines, giving top width, bottom width, wall angle and height to within about 10 nm [2]. Employing shorter wavelength illumination would therefore extend the use of coherence microscopy to future generations of design rules to below 100 nm.

Coherence microscopy is also well adapted for measuring the geometry of narrow, high aspect ratio vias and holes. These can be microns or tens of microns deep, and the depth can be measured as long as there is line of sight view (i.e. no significant overhang) and enough light reflected from the bottom surface to give a signal. The ability to 'see' the bottom also depends on the width of the opening; as the width approaches the lateral resolution of the system for high aspect ratio vias and holes, the intensity of the light reflected from the bottom decreases due to shadowing and diffraction from the edges. For via and hole widths approaching the optical resolution limit, the ability to profile the bottom surface then depends on the depth, the narrow opening decreasing both the angle of illumination and the angle of view.

Using the maximum intensity scan [8] the technique is also capable of producing qualitative views of surfaces with a very large depth of field — the restricted depth of field of the large numerical aperture is increased from a fraction of a micron to several microns or even tens of microns depending on the scan depth. This is useful when a qualitative appreciation is required, and provides a simpler solution than electron microscopy [10].

4. Conclusions

In this paper we have demonstrated that coherence microscopy is not limited to the use of broadband illumination sources to produce the narrow fringe envelope used as the optical 'probe', but that high intensity coherent UV sources can also be employed in combination with coherence reduction techniques such as that using the narrow depth of field of a large numerical aperture objective. Wavelength reductions can therefore be used to improve the lateral resolution for three dimensional profiling, leading to 100 nm or better lateral resolution and nanometric axial resolution. This means that inspection techniques can also follow the improvements in resolution of optical lithography, depending on the availability of light sources, optical materials for illumination and imaging and appropriate camera targets. UV coherence microscopy could therefore form the basis for a new generation of inspection tools for 300 mm and larger Si wafers with deep submicron features, being non-destructive, rapid and open to automation.

Acknowledgements

The authors would like to thank A. Golanski for helpful discussion concerning this work.

References

[1] The National Technology Roadmap for Semiconductors, SEMATECH, 1997 Edition.
[2] M. Davidson, K. Kaufman, I. Mazor, F. Cohen, SPIE 775 (1987) 233–247.
[3] P.J. Caber, S.J. Martinek, R.J. Niemann, SPIE 2088 (1993) 195–203.
[4] P. DeGroot, L. Deck, J. Mod. Optics 42 (1995) 389–401.
[5] P.C. Montgomery, C. Kazmierski, S. Bouchoule, Inst. Phys. Conf. Ser. No. 160 (1997) 51–54.
[6] J.P. Fillard, Near Field Optics and Nanoscopy, World Scientific, Singapore, 1996.
[7] M. Pluta, Advanced Light Microscopy, Vol. 1, Elsevier, Amsterdam, 1988.
[8] P.C. Montgomery, J.M. Lussert, P. Vabre, D. Benhaddou, SPIE 2412 (1995) 88–94.
[9] B.S. Lee, T.C. Strand, Appl. Optics 29 (26) (1990) 3784–3788.
[10] P.C. Montgomery, P. Gall-Borrut, Inst. Phys. Conf. Ser. No. 149 (1996) 147–152.